General mathematics for technicians
Second edition

H. G. Davies, M.Sc. (London), M.Inst.P.

Head of Department of Science
Carmarthen Technical and Agricultural College

G. A. Hicks, B.Sc. (Wales)

Senior Lecturer in Mechanical Engineering
Carmarthen Technical and Agricultural College

McGRAW-HILL Book Company (UK) Limited

London · New York · St. Louis · San Francisco
Auckland · Beirut · Bogotá · Düsseldorf · Johannesburg
Lisbon · Lucerne · Madrid · Mexico · Montreal
New Delhi · Panama · Paris · San Juan · São Paulo
Singapore · Sydney · Tokyo · Toronto

Published by
McGRAW-HILL Book Company (UK) Limited
MAIDENHEAD · BERKSHIRE · ENGLAND

British Library Cataloguing in Publication Data
Davies, Henri Gwyn
  General mathematics for technicians. – 2nd ed.
  1. Shop mathematics
  I. Title II. Hicks, Gordon Allen III. General mathematics for technical colleges
  510′.2′46   TJ1165   77–30172
ISBN 0–07–084222–1

PRINTED AND BOUND IN GREAT BRITAIN

In memory of Linda

Preface

This book, a revised edition of our previous title *General mathematics for technical colleges*, has been written and planned to meet the requirements of the mathematics Standard Unit Level One of the Technician Education Council's programmes in Engineering and Science.

At the beginning of each chapter a list of objectives is included. These objectives are the items of mathematics that a student is expected to understand and use after working through the chapter.

At the end of each chapter is an assessment test composed of short answer and multi-choice questions. These tests can be used to monitor a student's progress. Since these questions are more difficult to construct than the traditional questions, any comments regarding them will be greatly appreciated.

Assessment questions of this type only test one or two items of information. But a technician applying mathematics to technological problems must be able to handle several items of mathematical knowledge at the same time. To this end, short answer or multi-choice questions are not suitable. For this reason many traditional questions have also been included, both within each chapter and as a revision exercise at the end of the book. This revision exercise has been divided into sections.

The questions in the revision exercise have been selected from past Technician examination papers of the Regional Examining Boards. All such questions are used by kind permission of the following Boards:

East Midland Educational Union
Northern Counties Technical Examination Council
Union of Educational Institutes
Union of Lancashire and Cheshire Institutes
Welsh Joint Education Committee
Yorkshire and Humber Council for Further Education

Extracts have been included from Frank Castle's *Logarithm and other tables*, by kind permission of Macmillan Co. Ltd.

During the preparatory stages of the manuscript we were involved with the Dyfed Mathematics panel, which was convened to consider the standard units prepared by the Technician Education Council. It is a pleasure to record our indebtedness to the following members of the panel for many useful discussions: D. G. Hazelby, J. K. Jones, J. G. Thomas, and R. E. Warlow.

Finally, we would like to thank the editorial staff of McGraw-Hill for their forebearance and assistance over the whole period of the book's preparation.

H. G. Davies
G. A. Hicks

Contents

Preface

1. ARITHMETIC 1
   1.1 Mixed operations and the rules of precedence ... 1
   1.2 The three laws of arithmetic ... 3
   1.3 Factors and prime factors of numbers ... 4
   1.4 Square roots and cube roots using prime factors ... 5
   1.5 Highest Common Factor of two or more numbers (HCF) ... 7
   1.6 Lowest Common Multiple (LCM) of two or more numbers ... 8
   1.7 Vulgar fractions ... 8
   1.8 Comparison of fractions and addition and subtraction ... 10
   1.9 Multiplication and division of fractions ... 13
   1.10 Fractions involving a mixture of $+, -, \times \div$ ... 15
   1.11 Decimal factions ... 16
   1.12 Multiplying and dividing by 10, 100, 1000 ... 17
   1.13 Conversions of fractions ... 18
   1.14 Non-terminating decimal fractions and their reduction to a number of places ... 20
   1.15 Significant figures ... 21
   1.16 Basic operations with decimals ... 22

2. ARITHMETIC 2
   2.1 Ratio ... 29
   2.2 Proportion ... 32
   2.3 Percentages ... 35
   2.4 Indices ... 39
   2.5 Numbers expressed in standard form ... 43
   2.6 Binary numbers ... 45

3. CALCULATIONS
   3.1 Introduction ... 52
   3.2 Errors and accuracy ... 53
   3.3 Approximate values ... 54
   3.4 Aids to numerical calculations ... 55
   3.5 Evaluation of square roots, squares, and reciprocals by tables ... 55
   3.6 Logarithms ... 59
   3.7 Multiplication and division using logarithms ... 63
   3.8 Logarithms of numbers less than 1 ... 65

3.9	Logarithm of unity	67
3.10	Reciprocals using logarithm tables	67
3.11	Powers and roots	68
3.12	Powers and roots of numbers less than 1	69
3.13	The slide rule	71
3.14	Calculators	72
3.15	Comparison of aids	73

4. ALGEBRA 1—Basic operations

4.1	Letters, and their addition and subtraction	78
4.2	Three laws of algebra	80
4.3	Simple multiplication and division	82
4.4	The rules of precedence	83
4.5	Simple substitution	83
4.6	Directed numbers	84
4.7	Indices	90
4.8	Brackets	95
4.9	Multiplication of binomials	97
4.10	Factors	97

5. ALGEBRA 2

5.1	Expressions and equations	105
5.2	Simple equations	106
5.3	More difficult simple equations	108
5.4	Construction and solution of equations in engineering and science	110
5.5	Simultaneous equations	112
5.6	Evaluation of formulae	114
5.7	Transposition of formulae	116

6. DIAGRAMS AND GRAPHS

6.1	Conversion of data into a related system of units	124
6.2	Representation of related values using two parallel axes	125
6.3	Mappings	129
6.4	Two axes at right angles	129
6.5	Scales on the axes	130
6.6	Plotting a point given its co-ordinates	132
6.7	Straight-line graphs	133
6.8	Gradient of a straight-line graph	134
6.9	Values from a straight-line graph	134

7. STATISTICS
    7.1 Introduction   140
    7.2 Display of data   140
    7.3 Frequency table   141
    7.4 Pictorial displays   143

8. GEOMETRY 1
    8.1 Angles   156
    8.2 Properties of angles   157
    8.3 Angles of a triangle   162
    8.4 Pythagoras' theorem   166
    8.5 Construction of a right angle   168
    8.6 Congruent triangles   170
    8.7 Similar triangles   173
    8.8 Construction of triangles   176

9. GEOMETRY 2
    9.1 The circle   187
    9.2 Angles subtended by an arc of a circle   190
    9.3 Properties of tangents   193
    9.4 Property of a chord   194
    9.5 Radian measure and length of arc   196
    9.6 Quadrilaterals   200
    9.7 Regular hexagon   202

10. AREA AND VOLUME
    10.1 Perimeter and area   209
    10.2 Areas of plane rectilinear figures   210
    10.3 Areas of circular figures   217
    10.4 Volumes of prisms   221
    10.5 Volumes and surface areas of prisms   222
    10.6 Volume and surface area of a cylinder   226
    10.7 Volume of a sphere   229
    10.8 Volume of a pyramid and a cone   231

11. TRIGONOMETRY
    11.1 Introduction   241
    11.2 Trigonometrical ratios of acute angles   242
    11.3 Use of trigonometrical tables   243
    11.4 Trigonometrical ratios for 30°, 45°, and 60°   246
    11.5 Sine and cosine of complementary angles   248
    11.6 Solution of right-angled triangles   248

11.7 Trigonometrical graphs   254
11.8 Sine and cosine waves   256

Revision exercise   262
Answers   281
Index   293

1. Arithmetic 1

Objectives

After working through this chapter you should be able to

1. Apply the rules of precedence to numbers and fractions.
2. Apply the commutative, associative, and distributive laws.
3. Find the prime factors of numbers.
4. Calculate the square and cube roots of numbers using prime factors.
5. Determine the Highest Common Factor of two or more numbers.
6. Determine the Lowest Common Multiple of two or more numbers.
7. Simplify, add, subtract, multiply, and divide vulgar fractions.
8. Recognize proper, improper, and mixed fractions.
9. Multiply decimal numbers by multiples of 10.
10. Divide decimal numbers by multiples of 10.
11. Convert vulgar fractions to decimals.
12. Convert decimals to vulgar fractions.
13. Distinguish between a terminating and non-terminating decimal.
14. Define a recurring decimal.
15. Reduce a number to a given number of decimal places.
16. Reduce a number to a given number of significant figures.
17. Add, subtract, multiply, and divide decimal numbers.

1.1 Mixed operations and the rules of precedence

The word 'operation' means $+$, $-$, $\times$ or $\div$. When a mixture of operations are used in a calculation it is necessary to decide in which order they are carried out. The rules which determine this order are called the 'rules of precedence'. In the following simple calculation

$$3 + 4 \times 5 - 10$$

the order in which it is to be worked out must be decided. The order is decided in accordance with Rule 1.

Rule 1. $\times$ *and* $\div$ *is carried out before* $+$ *and* $-$.

Hence in the above calculation, we have

$$3 + 20 - 10 = 13.$$

Again in
$$63 \div 7 - 4 \times 2$$
the $\div$ and $\times$ are carried out first to give
$$9 - 8 = 1.$$
Sometimes a calculation contains a bracket. In this case Rule 2 is used.

Rule 2. *Calculations inside the bracket are always carried out first.*

For example, in the calculation
$$63 \div (7 - 4) \times 2$$
the inside of the bracket must be worked out first, to give
$$63 \div 3 \times 2 = 21 \times 2 = 42$$

Rules 1 and 2 may be summarized by the word BODMAS, which identifies the order of preference for *B*rackets, *O*f, *D*ivision, *M*ultiplication, *A*ddition, *S*ubtraction.

EXAMPLE 1.1   Evaluate

(a) $30 \div (10 - 4) + 2$

(b) $30 \div 10 - 4 + 2$

(a) $30 \div (10 - 4) + 2 = 30 \div 6 + 2,$ using Rule 2
$$= 5 + 2, \quad \text{using Rule 1}$$
$$= 7$$

(b) $30 \div 10 - 4 + 2 = 3 - 4 + 2,$ using Rule 1
$$= 1$$

EXERCISE 1.1

Evaluate the following:

1. $6 \times 2 - 3$ 6. $6 \times (4 - 2)$

2. $27 \div 3 - 27 \div 9$ 7. $(27 - 20) \times (27 + 3)$

3. $14 + 2 \times 3 - 28 \div 2$ 8. $(14 + 2 \times 3 - 10) \div 2$

4. $6 - 2 \times 3 + 10$ 9. $(6 - 2) \times 3 + 10$

5. $10 \times 5 - 3 \times 5$ 10. $(3 + 5) \times \{3 + (6 - 4)\}$

Note: The two rules of precedence apply to vulgar and decimal fractions, as well as simple numbers.

1.2 The three laws of arithmetic

(a) Commutative law

Consider the operations

$$6+4 = 4+6$$

and

$$3\times5 = 5\times3$$

It does not matter which of the figures is written down first, the answer is the same. These two results constitute the commutative law, which states that in the addition or multiplication of two numbers the order in which they are written down is immaterial.

(b) Associative law

In arithmetic the addition or multiplication of three numbers is independent of the order in which the operation is carried out. For example, it can easily be seen that

$$4+(7+3) = (4+7)+3$$

that is

$$4+ \quad 10 \quad - \quad 11 \quad +3$$

Again

$$3\times(4\times5) - (3\times4)\times5$$

since

$$3\times \quad 20 \quad = \quad 12 \quad \times5$$

The two results constitute the associative law.

(c) Distributive law

The following calculation is an example of the distributive law

$$5\times(2+4) = 5\times2+5\times4$$

The result is readily seen to be true since

$$5\times6 = \quad 10 \quad + \quad 20$$

Again

$$7\times(9-4) = 7\times9-7\times4$$

since

$$7\times \quad 5 \quad = \quad 63 \quad - \quad 28$$

EXERCISE 1.2

Verify the following, and state which law each obeys.

1. $5+3 = 3+5$ 7. $7+(3+2) = (7+3)+2$

2. $4 \times 7 = 7 \times 4$ 8. $9 \times (4 \times 5) = (9 \times 4) \times 5$

3. $8+7 = 7+8$ 9. $3 \times (8+2) = 3 \times 8 + 3 \times 2$

4. $3 \times 1 = 1 \times 3$ 10. $5 \times (5+6) = 5 \times 5 + 5 \times 6$

5. $8+(2+3) = (8+2)+3$ 11. $8 \times (8-3) = 8 \times 8 - 8 \times 3$

6. $4 \times (6 \times 2) = (4 \times 6) \times 2$ 12. $6 \times (7-5) = 6 \times 7 - 6 \times 5$

1.3 Factors and prime factors of numbers

Consider the number 42. The number 6 will divide exactly into it, that is:

$$42 = 6 \times 7$$

6 and 7 are called **factors** of 42. A factor of any number is any other number which will divide exactly into it.

Some factors can themselves have factors. For example, 6 has two factors, 3 and 2. Other factors, such as 7 do not have factors. Such factors are called **prime factors**. A prime factor cannot be expressed in further factors. In the example above, 42 has two factors 6 and 7, but its prime factors are 2, 3, and 7. The **prime numbers** therefore are those numbers which have no factors; they are:

$$1, 2, 3, 5, 7, 11, \ldots, \text{etc.}$$

The prime factors of any number may be obtained by

(a) dividing repeatedly by 2 until 2 ceases to be a factor,

(b) dividing repeatedly by 3 until 3 ceases to be a factor and so on with 5, 7, 11, etc. The method is shown in Example 1.2.

EXAMPLE 1.2   Determine the prime factors of 720.

```
2 | 720 ┐
2 | 360 │  divide repeatedly by 2
2 | 180 │
2 |  90 ┘
3 |  45     2 ceases to be a factor—try 3
3 |  15
5 |   5     3 ceases to be a factor—try 5
  |   1
```

Therefore the prime factors of 720 are $2 \times 2 \times 2 \times 2 \times 3 \times 3 \times 5$.

It is worth knowing the following facts when finding prime factors.

(a) A number is exactly divisible by 2 if the last digit on the right is 0 or an even number, e.g., 13$\overset{*}{6}$ or 72$\overset{*}{0}$.

(b) A number is exactly divisible by 3 if the sum of the digits is divisible by 3; e.g., 51 is divisible by 3 since $5+1 = 6$ which is divisible by 3.

(c) A number is exactly divisible by 5 if the last digit on the right is 0 or 5, e.g., 42$\overset{*}{0}$ or 14$\overset{*}{5}$.

EXERCISE 1.3

1. State whether 2, 3 or 5 is a factor of the following:
   (a) 202   (b) 550   (c) 363   (d) 300   (e) 729

2. State which of the following are prime numbers:
   3, 4, 5, 9, 10, 11, 13, 15, 16, 17

3. Find the prime factors of the following:
   (a) 48   (b) 76   (c) 128   (d) 920   (e) 108   (f) 210   (g) 525

4. Find the prime factors of:
   (a) 18 816   (b) 111 111   (c) 131 313

1.4   Square roots and cube roots using prime factors

(a) Square roots

Consider a number such as 9. Its square root is defined as that number, which multiplied by itself, gives an answer of 9. Such a number is 3, since $3 \times 3 - 9$. Therefore, the square root of 9 is 3, and is written as

$$\sqrt{9} = 3$$

Again

$$25 = 5 \times 5$$

so that

$$\sqrt{25} = 5$$

Thus whenever a number can be expressed in terms of a pair of identical factors, the square root will be one of these factors.

Most numbers do not have exact square roots, numbers such as 2, 5, 7, etc. Numbers such as 4, 9, 25, 36, which have exact square roots are called **perfect squares**.

The square roots of perfect squares can be found using *prime* factors, as follows.

Step 1. Express the number in prime factors.

5

Step 2. Select one from each pair of factors.

Step 3. Multiply out the prime factors selected to obtain the square root.

EXAMPLE 1.3   Find the square root of 400.

Step 1. $$400 = 2 \times 2 \times 2 \times 2 \times 5 \times 5$$

Step 2. $$2 \qquad 2 \qquad 5$$

Step 3. $$\sqrt{400} = 2 \times 2 \times 5 = 20$$

A number such as 800 is seen *not* to be a perfect square if it is expressed in prime factors.

$$800 = 2 \times 2 \times 2 \times 2 \times \overset{*}{2} \times 5 \times 5$$

This can be deduced since there is one factor 2 (shown with an asterisk) which is not paired. In order to make 800 into a perfect square it must be multiplied by 2 to make up a pair. Therefore to convert 800 into a perfect square it must be multiplied by 2.

$$\text{Perfect square} = 800 \times 2 \qquad = 1600$$
$$\text{Square root} \quad = 2 \times 2 \times 2 \times 5 = \quad 40$$

(b) Cube roots

The cube root of a number may be found using a similar method. If the number can be expressed as a trio of identical factors the cube root will be one of these factors, for example,

$$8 = 2 \times 2 \times 2$$

then

$$\sqrt[3]{8} = \quad 2$$

A number such as 8 is called a **perfect cube**. The cube roots of such numbers can be obtained as follows.

Step 1. Write the numbers in prime factors.

Step 2. Select one from each trio of factors.

Step 3. Multiply out the factors selected to obtain the cube root.

EXAMPLE 1.4   Determine the cube root of 3375

Step 1. Prime factors: $3375 = 3 \times 3 \times 3 \times 5 \times 5 \times 5$

Step 2. Select one from each trio: $3 \quad \times \quad 5$

Step 3. Multiply: $\sqrt[3]{3375} = \qquad 15$

6

EXERCISE 1.4

1. Determine the square roots of the following
   (a) 196 (b) 900 (c) 144 (d) 784 (e) 3969 (f) 63 504 (g) 1764
   (h) 11 025

2. Find the least number which will make the following (a) perfect squares,
   (b) perfect cubes
   (a) 54 (b) 24 (c) 50 (d) 72 (e) 108 (f) 98 (g) 200 (h) 500

3. Determine the cube roots of the following
   (a) 729 (b) 216 (c) 64 (d) 1728 (e) 2744 (f) 512 (g) 1000 (h) 9261

1.5 Highest Common Factor of two or more numbers (HCF)

Consider a set of two or more numbers. The HCF is the biggest number
which will divide exactly into the whole set. For example, for the two numbers
8 and 12, the biggest number that will divide exactly into both is 4. Hence,
their HCF is 4.

The method of determining the HCF of a set of numbers is as follows:

Step 1. Write each member of the set in prime factors.

Step 2. Select all factors that are common to the whole set

Step 3. The product of the selected factors is the HCF.

EXAMPLE 1.5   Find the HCF of 180, 600, and 1260.

Step 1.
$$180 = 2 \times 2 \times 3 \times 3 \times 5$$
$$600 = 2 \times 2 \times 2 \times 3 \times 5 \times 5$$
$$1260 = 2 \times 2 \times 3 \times 3 \times 5 \times 7$$

Step 2. Select common factors shown between dotted lines.

Step 3.   HCF $= 2 \times 2 \times 3 \times 5$

$$= \quad 60$$

EXERCISE 1.5

Determine the HCF of the numbers in each question.

1. 60, 90, 150   6. 18, 36, 81
2. 60, 72, 84   7. 420, 90, 210
3. 12, 18, 24, 36   8. 525, 315, 1575
4. 30, 40, 120   9. 68, 184, 96
5. 32, 72, 96   10. 15, 50, 35

1.6 Lowest Common Multiple (LCM) of two or more numbers

In a set of two or more numbers the LCM is the smallest number into which each member of the set will divide exactly. For example, the LCM of 6 and 8 is 24.

The LCM of a set of numbers can be found as follows.

Step 1. Write each number in prime factors.

Step 2. Choose the most factors of any kind that occur in any number.

Step 3. The LCM is the product of the factors chosen.

EXAMPLE 1.6   Find the LCM of 32, 48, 72.

Step 1. $32 = 2 \times 2 \times 2 \times 2 \times 2$
$48 = 2 \times 2 \times 2 \times 2 \quad \times 3$
$72 = 2 \times 2 \times 2 \quad\quad \times 3 \times 3$

Step 2. The greatest number of 2's is in 32, that is, $2 \times 2 \times 2 \times 2 \times 2$.
The greatest number of 3's is in 72, that is, 3×3.

Step 3. LCM $= 2 \times 2 \times 2 \times 2 \times 2 \times 3 \times 3$
$ = 288$

EXERCISE 1.6

Determine the LCM of the numbers in each question.

1. 12, 15, 18
2. 15, 25, 45
3. 27, 54, 81
4. 35, 42, 98
5. 105, 225, 490

6. 22, 33, 55
7. 36, 63, 84
8. 28, 30, 315
9. 252, 140, 315
10. 77, 231, 70, 165

1.7 Vulgar fractions

This is the name given to proper, improper and mixed fractions.

(a) Proper fractions

Consider a paved area outside a house composed of eight paving slabs of equal size, as shown in Fig. 1.1.

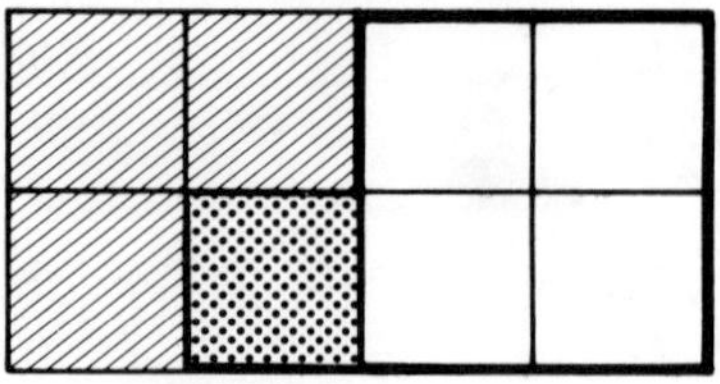

Fig. 1.1

Let three slabs be pink as shown by the line shading,

one slab be green as shown by the dotted shading,

and four slabs be grey, shown unshaded.

Three parts out of eight are coloured pink, that is, $\frac{3}{8}$ of the whole area.

In the fraction $\frac{3}{8}$ the top number 3 is called the **numerator** and the bottom number 8 is called the **denominator**. Any fraction, where the numerator is smaller than the denominator is called a **proper fraction**.

(b) Simplification of fractions

Four parts out of eight are coloured grey, that is, $\frac{4}{8}$ of the whole area. From Fig. 1.1 however it is seen that the grey part (unshaded) forms $\frac{1}{2}$ the whole area, that is,

$$\tfrac{4}{8} = \tfrac{1}{2}$$

The above result can be obtained by dividing the numerator and denominator by 4. The process is known as **simplification**, and the procedure is called **cancelling**. Any fraction may be simplified in this way if both numerator and denominator are divisible by the same number.

In the fraction $\frac{6}{15}$, 3 will divide exactly into the numerator and denominator, so that

$$\frac{\cancel{6}^{\,2}}{\cancel{15}_{\,5}} = \frac{2}{5}$$

(c) Mixed and improper fractions

A mixed fraction contains a whole number and a fraction, for example, $2\frac{3}{4}$.

An improper fraction has its numerator greater than its denominator, for example, $\frac{13}{6}$.

A mixed fraction can be converted into an improper fraction, and vice versa, as shown in the following:

$$2\tfrac{3}{4} = 2 \times 4 \text{ quarters plus 3 quarters} = \tfrac{11}{4}$$
$$\tfrac{13}{6} = 2 \text{ whole numbers plus } \tfrac{1}{6} \quad = 2\tfrac{1}{6}$$

EXERCISE 1.7

1. In the following state the numerator and denominator in each case.

   (a) $\frac{1}{7}$ (b) $\frac{2}{5}$ (c) $\frac{3}{4}$ (d) $\frac{18}{19}$ (e) $\frac{19}{3}$ (f) $\frac{21}{4}$

2. Fill the missing blanks in the following:

   (a) $\dfrac{4}{12} = \dfrac{*}{3}$ (b) $\dfrac{8}{12} = \dfrac{2}{*}$ (c) $\dfrac{15}{6} = \dfrac{*}{2}$

(d) $\frac{7}{49} = \frac{1}{*}$ (e) $\frac{8}{20} = \frac{2}{*}$ (f) $\frac{2}{3} = \frac{8}{*}$

(g) $\frac{1}{5} = \frac{10}{*}$ (h) $\frac{6}{1} = \frac{*}{5}$ (i) $7 = \frac{*}{3}$

(j) $8 = \frac{16}{*}$

3. Simplify
   (a) $\frac{13}{39}$ (b) $\frac{10}{25}$ (c) $\frac{14}{35}$ (d) $\frac{44}{33}$ (e) $\frac{15}{3}$

4. Convert into improper fractions
   (a) $2\frac{3}{4}$ (b) $3\frac{3}{8}$ (c) $4\frac{5}{21}$ (d) $4\frac{7}{10}$ (e) $1\frac{5}{6}$
   (f) $10\frac{7}{10}$ (g) $11\frac{10}{11}$ (h) 3 (i) 5 (j) 10

5. Convert into mixed fractions
   (a) $\frac{14}{9}$ (b) $\frac{21}{10}$ (c) $\frac{25}{7}$ (d) $\frac{11}{6}$ (e) $\frac{100}{9}$
   (f) $\frac{18}{10}$ (g) $\frac{19}{13}$ (h) $\frac{26}{13}$ (i) $\frac{8}{4}$
   (j) $\frac{7}{1}$ (k) $\frac{121}{1}$

1.8 Comparison of fractions and addition and subtraction

(a) Comparison of fractions

In order to compare the sizes of fractions it is necessary to write them all with the same denominator. For example, with the two fractions $\frac{1}{2}$ and $\frac{2}{5}$ we must use a denominator into which both 2 and 5 will divide exactly. Such a number, of course, is the LCM, which is 10.

$$\frac{1}{2} = \frac{*}{10} = \frac{5}{10}$$

$$\frac{2}{5} = \frac{*}{10} = \frac{4}{10}$$

Therefore $\frac{1}{2}$ is the bigger of the two.

(b) Addition of fractions

As in (a) it is necessary to re-write the fractions, all with the same denominator, in order to add them together.

10

EXAMPLE 1.7 Add the fractions $\frac{1}{4}+\frac{3}{5}+\frac{1}{6}$.

$$\text{LCM of } 4, 5, 6 = 60$$

$$\frac{1}{4}+\frac{3}{5}+\frac{1}{6} = \frac{*}{60}+\frac{*}{60}+\frac{*}{60}$$

$$= \frac{15}{60}+\frac{36}{60}+\frac{10}{60}$$

$$= \frac{61}{60}$$

$$= 1\frac{1}{60}$$

If the fractions are mixed the whole numbers are added separately, and then the proper fractions, as shown in Example 1.8.

EXAMPLE 1.8 Evaluate $2\frac{1}{4}+3\frac{2}{3}+1\frac{1}{2}$.

Adding the whole numbers separately gives

$$6+\frac{1}{4}+\frac{2}{3}+\frac{1}{2} = 6+\frac{3}{12}+\frac{8}{12}+\frac{6}{12}, \quad \text{since LCM of } 4, 3, 2 = 12$$

$$= 6+\frac{17}{12}$$

$$= 6+1\frac{5}{12}$$

$$= 7\frac{5}{12}$$

(c) Subtraction of fractions

The first step in the subtraction of proper fractions is to find the LCM of the denominators, as shown in Example 1.9 (a).

For mixed fractions the whole numbers are again worked out separately, as in Example 1.9 (b).

EXAMPLE 1.9 Subtract

(a) $\frac{7}{10}-\frac{4}{15}$

(b) $3\frac{3}{5}-1\frac{1}{6}$

(a) LCM of 10 and 15 = 30
　　　Therefore

$$\frac{7}{10}-\frac{4}{15} = \frac{21}{30}-\frac{8}{30}$$

$$= \frac{13}{30}$$

11

(b) First, the whole numbers are subtracted, to give

$$3\tfrac{3}{5}-1\tfrac{1}{6} = 2+\tfrac{3}{5}-\tfrac{1}{6}$$

$$= 2+\tfrac{18}{30}-\tfrac{5}{30}, \quad \text{since LCM of 5 and 6} = 30$$

$$= 2+\tfrac{13}{30}$$

$$= 2\tfrac{13}{30}$$

One complication arises in the subtraction of mixed fractions, when the proper fraction being subtracted is bigger than the other proper fraction, as in Example 1.10.

EXAMPLE 1.10 Evaluate $5\tfrac{1}{7}-2\tfrac{2}{5}$.

Proceeding as in Example 1.11(b) we have

$$5\tfrac{1}{7}-2\tfrac{2}{5} = 3+\tfrac{1}{7}-\tfrac{2}{5}$$

$$= 3+\tfrac{5}{35}-\tfrac{14}{35} \quad \text{since the LCM of 7 and 5} = 35$$

14 cannot be subtracted from 5. Therefore, one of the whole numbers is taken and converted into a fraction with 35 as denominator, to give

$$2+\frac{35}{35}+\frac{5}{35}-\frac{14}{35} = 2+\frac{40-14}{35}$$

$$= 2\tfrac{26}{35}$$

A mixture of addition and subtraction can be worked out in this way as shown in Example 1.11.

EXAMPLE 1.11 Evaluate $4\tfrac{1}{5}+2\tfrac{1}{10}-3\tfrac{8}{15}$.

$$4\tfrac{1}{5}+2\tfrac{1}{10}-3\tfrac{8}{15} = 3+\tfrac{1}{5}+\tfrac{1}{10}-\tfrac{8}{15}$$

$$= 3+\frac{*\ +\ *\ -\ *}{30}, \quad \text{since LCM} = 30$$

$$= 3+\frac{6+3-16}{30}$$

$$= 3+\frac{9-16}{30}$$

$$= 2+\frac{30}{30}+\frac{9-16}{30}$$

$$= 2+\frac{39-16}{30}$$

$$= 2\tfrac{23}{30}$$

EXERCISE 1.8

1. By re-writing the fractions in each question with the same denominator place the fractions in correct order of size, starting with the biggest.

(a) $\frac{1}{6}, \frac{2}{5}, \frac{2}{15}$ (b) $\frac{3}{8}, \frac{5}{12}, \frac{1}{3}$ (c) $\frac{2}{3}, \frac{1}{4}, \frac{7}{12}$

(d) $\frac{3}{4}, \frac{4}{7}, \frac{9}{14}$ (e) $\frac{7}{25}, \frac{3}{10}, \frac{4}{5}$

2. Evaluate the following:

(a) $\frac{3}{7} + \frac{2}{5}$ (b) $\frac{3}{4} + \frac{2}{3}$ (c) $\frac{4}{5} + \frac{3}{10} + \frac{2}{20}$

(d) $\frac{3}{11} + \frac{3}{22} + \frac{4}{33}$ (e) $2\frac{1}{4} + 4\frac{2}{9}$ (f) $1\frac{7}{16} + 2\frac{7}{20}$

(g) $3\frac{3}{21} + 2\frac{4}{15}$ (h) $2\frac{1}{4} + 3\frac{1}{5} + 4\frac{1}{10}$ (i) $4\frac{1}{9} + 2\frac{1}{3} + 1\frac{7}{12}$

3. Carry out the following subtractions:

(a) $\frac{7}{16} - \frac{3}{8}$ (b) $\frac{4}{5} - \frac{7}{25}$ (c) $\frac{5}{6} - \frac{7}{9}$

(d) $2\frac{2}{3} - 1\frac{1}{4}$ (e) $4\frac{7}{15} - 1\frac{3}{20}$ (f) $3\frac{2}{11} - 1\frac{3}{44}$

(g) $3\frac{1}{4} - 1\frac{4}{5}$ (h) $6\frac{7}{16} - 2\frac{23}{24}$ (i) $2\frac{3}{4} - 1\frac{7}{8}$

4. Work out the following:

(a) $2\frac{3}{4} - 1\frac{7}{16} + 2\frac{1}{8}$

(b) $2\frac{3}{8} + 1\frac{4}{15} - 3\frac{3}{20}$

(c) $1\frac{2}{3} + 4\frac{7}{12} - 2\frac{5}{6}$

(d) $3\frac{1}{4} - 2\frac{9}{3} + 1\frac{7}{24}$

(e) $6\frac{3}{10} - 2\frac{4}{25} - 1\frac{2}{5}$

1.9 Multiplication and division of fractions

(a) Multiplication

Two fractions are multiplied together by multiplying the two numerators and the two denominators, as follows

$$\frac{2}{3} \times \frac{4}{5} = \frac{2 \times 4}{3 \times 5} = \frac{8}{15}$$

A fraction *of* another fraction is treated as a multiplication, that is

$$\frac{2}{5} \text{ of } \frac{2}{3} = \frac{2}{5} \times \frac{2}{3} = \frac{4}{15}$$

Whenever possible the multiplication should first be simplified by cancelling, as shown in Example 1.12.

EXAMPLE 1.12   Find $\frac{5}{9} \times \frac{12}{25}$.

It is obvious that both 3 and 5 will divide into numerators and denominators. The fractions are simplified by cancelling.

$$\frac{\cancel{5}^{1}}{\cancel{9}_{3}} \times \frac{\cancel{12}^{4}}{\cancel{25}_{5}} = \frac{1 \times 4}{3 \times 5} = \frac{4}{15}$$

The multiplication of mixed fractions can be carried out in the same way, if, first of all, they are changed into improper form. The procedure is then exactly the same as in Example 1.12.

EXAMPLE 1.13   Evaluate $2\frac{1}{5} \times 2\frac{1}{7}$.

$$2\frac{1}{5} \times 2\frac{1}{7} = \frac{11}{\cancel{5}_{1}} \times \frac{\cancel{15}^{3}}{7}$$

$$= \frac{11 \times 3}{1 \times 7}$$

$$= \frac{33}{7}$$

$$= 4\frac{5}{7}$$

(b) Division of fractions

The method of dividing fractions can be explained using the following simple calculations.

If we divide eight apples between two persons each receives four. Thus $8 \div 2$ finds how many pairs there are in 8.

Similarly, $8 \div \frac{1}{2}$ is the number of halves in $8 = 16$

$8 \div \frac{1}{4}$ is the number of quarters in $8 = 32$

Again $6 \div 1\frac{1}{2}$ is the number of $1\frac{1}{2}$ in $6 = 4$

The above results can be obtained as follows:

$$8 \div 2 = \frac{8}{1} \div \frac{2}{1} = \frac{8}{1} \times \frac{1}{2} = 4$$

$$8 \div \frac{1}{2} = \frac{8}{1} \div \frac{1}{2} = \frac{8}{1} \times \frac{2}{1} = 16$$

$$6 \div 1\frac{1}{2} = \frac{6}{1} \div \frac{3}{2} = \frac{\cancel{6}^{2}}{1} \times \frac{2}{\cancel{3}_{1}} = 4$$

The rule is
(a) make both fractions improper,

(b) change $\div$ to $\times$, and at the same time invert the fraction on the right of $\div$.

EXAMPLE 1.14   Simplify $3\frac{3}{5} \div 2\frac{7}{10}$.

$$3\frac{3}{5} \div 2\frac{7}{10} = \frac{18}{5} \div \frac{27}{10}$$

$$= \frac{{}^2\cancel{18}}{{}_1\,5} \times \frac{\cancel{10}\,{}^2}{\cancel{27}\,{}_3}$$

$$= \frac{4}{3}$$

$$= 1\frac{1}{3}$$

EXERCISE 1.9

Evaluate the following:

1. $\frac{7}{12} \times \frac{3}{14}$ 2. $\frac{9}{13} \times \frac{26}{27}$ 3. $\frac{3}{4} \times \frac{2}{9}$

4. $2\frac{3}{4} \times 1\frac{1}{11}$ 5. $3\frac{1}{8} \times 2\frac{2}{5}$ 6. $4\frac{2}{7} \times 5\frac{5}{6}$

7. $\frac{3}{5} \div \frac{7}{10}$ 8. $\frac{2}{3} \div \frac{5}{6}$ 9. $\frac{5}{7} \div \frac{10}{21}$

10. $5\frac{1}{3} \div 2\frac{2}{9}$ 11. $1\frac{2}{3} \div 1\frac{1}{4}$ 12. $4\frac{2}{5} \div 1\frac{1}{10}$

13. $3\frac{3}{5} \div 3\frac{3}{4} \times 4\frac{1}{6}$ 14. $4\frac{5}{7} \times 1\frac{4}{11} \div 1\frac{9}{21}$ 15. $1\frac{5}{16} \div 2\frac{4}{5} \div 6\frac{1}{4}$

16. $22\frac{1}{2} \div 2\frac{4}{7} \div 3\frac{1}{5}$ 17. $3\frac{1}{3} \times 1\frac{5}{9} \times 3\frac{3}{20}$

1.10   Fractions involving a mixture of $+$, $-$, $\times$, $\div$

The laws of precedence stated in Section 1.1 apply to fractions, namely,

(a) inside of brackets are worked out first,

(b) $\times$ and $\div$ are carried out before $+$ and $-$

EXAMPLE 1.15   Simplify:

(a) $1\frac{3}{4} - 1\frac{2}{5} \times 1\frac{1}{14}$

(b) $(1\frac{3}{4} - 1\frac{2}{5}) \times 1\frac{1}{14}$

(a) $\times$ is carried out first:

$$1\frac{3}{4} - 1\frac{2}{5} \times 1\frac{1}{14} = 1\frac{3}{4} - \frac{\cancel{7}}{5} \times \frac{\cancel{15}\,{}^3}{\cancel{14}\,{}_2}$$

$$= 1\frac{3}{4} - 1\frac{1}{2}$$

$$= \frac{1}{4}$$

15

(b) Brackets evaluated first:

$$(1\tfrac{3}{4} - 1\tfrac{2}{5}) \times 1\tfrac{1}{14} = \left(\frac{15-8}{20}\right) \times 1\tfrac{1}{14}$$

$$= \tfrac{7}{20} \times 1\tfrac{1}{14}$$

$$= \frac{\overset{1}{7}}{\underset{4}{20}} \times \frac{\overset{3}{15}}{\underset{2}{14}}$$

$$= \tfrac{3}{8}$$

EXERCISE 1.10

Simplify the following:

1. $(\tfrac{3}{4} + \tfrac{1}{2}) \times \tfrac{2}{5}$
2. $\tfrac{3}{4} + \tfrac{1}{2} \times \tfrac{2}{5}$
3. $\tfrac{7}{12} \div \tfrac{3}{4} - \tfrac{7}{18}$
4. $\tfrac{7}{12} \div (\tfrac{3}{4} - \tfrac{7}{18})$
5. $2\tfrac{1}{4} \times 1\tfrac{2}{3} + 3\tfrac{4}{7} \times 1\tfrac{13}{15}$
6. $4\tfrac{4}{9} \times (1\tfrac{2}{5} + 1\tfrac{1}{4}) \div 2\tfrac{7}{9}$
7. $\dfrac{1\tfrac{3}{4} + 1\tfrac{3}{8}}{3\tfrac{5}{12} - 1\tfrac{1}{3}}$

1.11 Decimal fractions

A number such as 476 is composed of 4 hundreds, 7 tens, 6 units, that is, $400 + 70 + 6$. This system is extended to quantities less than 1, where they are expressed in tenths, hundredths, etc. Quantities less than 1 expressed in this way are **decimal fractions**, and are separated from numbers greater than 1 by a decimal point; for example, 476·38 consists of the whole number part and the decimal fraction part 0·38.

Note: If the number is less than 1 it is always written with a 0 before the decimal point, e.g. 0·59.

The number 476·38 is composed of

$$4 \times 100 = 400\cdot$$
$$7 \times 10 \; = \; 70\cdot$$
$$6 \times 1 \; \; = \; \; 6\cdot$$
$$3 \times \tfrac{1}{10} \; = \; \; 0\cdot3$$
$$8 \times \tfrac{1}{100} = \; \; 0\cdot08$$

Therefore, the decimal fraction 0·38 is

$$\tfrac{3}{10} + \tfrac{8}{100} = \tfrac{38}{100}$$

Note: The difference between, say, 0·3 and 0·03 must be clearly understood,

$$0·3 = \tfrac{3}{10}$$
$$0·03 = \tfrac{3}{100}$$

EXERCISE 1.11

Using graph paper with one large square = one whole number

and one small square = one hundredth

illustrate the following numbers by 'shading in' areas.

1. 0·6   2. 0·4   3. 0·7   4. 0·06   5. 0·02
6. 0·77 7. 0·86 8. 0·92 9. 0·1   10. 0·01
11. 3·2   12. 2·4   13. 7·53 14. 1·72 15. 2·03

1.12 Multiplying and dividing by 10, 100, 1000

If a whole number is multiplied by 10, one 0 is added on the end of the number. If it is multiplied by 100, two 0's are added, and by 1000 three 0's are added,

$$4 \times 10 = 40$$
$$23 \times 100 = 2300$$
$$14 \times 1000 = 14\ 000$$

In a whole number the decimal point is situated on the extreme right, so that in the above examples multiplying has the effect of moving the decimal point to the right. This applies to any decimal number, as shown:

× 10: decimal point moved one place to the right

$$21·45 \times 10 = 214·5$$

× 100: decimal point moved two places to the right

$$34·62 \times 100 = 3462$$

× 1000: decimal point moved three places to the right

$$0·176 \times 1000 = 176$$

Dividing is the reverse, as follows:

÷ 10: decimal point moved one place to the left

$$34·78 \div 10 = 3·478$$

÷ 100: decimal point moved two places to the left

$$247· \div 100 = 2·47$$

÷ 1000: decimal point moved three places to the left

$$3·4 \div 1000 = 0·0034$$

EXERCISE 1.12

Evaluate the following:

1. $14 \cdot 76 \times 10$ 2. 136×100 3. $0 \cdot 0071 \times 100$
4. $0 \cdot 7123 \times 1000$ 5. $14 \cdot 712 \times 100$ 6. $0 \cdot 7 \times 1000$
7. $0 \cdot 0214 \times 100$ 8. $101 \cdot 0101 \times 100$ 9. $1 \cdot 6 \times 100$
10. $0 \cdot 007172 \times 10\,000$ 11. $400 \div 10$ 12. $400 \div 1000$
13. $2 \cdot 712 \div 100$ 14. $612 \div 100$ 15. $716 \div 1000$
16. $0 \cdot 61 \div 100$ 17. $0 \cdot 0602 \div 100$ 18. $146 \cdot 13 \div 10\,000$
19. $1 \cdot 617 \div 10$ 20. $71 \cdot 6 \div 1000$

1.13 Conversions of fractions

(a) Decimals to vulgar fractions

A decimal fraction such as $0 \cdot 3$ as seen in Section 1.12 is 3 parts out of 10, so that

$$0 \cdot 3 = \tfrac{3}{10}$$

In the same section it is seen that $0 \cdot 38$ is 38 parts out of 100, so that

$$0 \cdot 38 = \tfrac{38}{100}$$

It follows that $0 \cdot 004$ is 4 parts out of 1000, that is $\tfrac{4}{1000}$, and $0 \cdot 617$ is 617 parts out of 1000, that is $\tfrac{617}{1000}$.

If the decimal fraction contains whole numbers then so will the vulgar fraction,

$$3 \cdot 47 = 3\tfrac{47}{100}$$

The vulgar fractions can be simplified by cancelling, as shown in Example 1.16.

EXAMPLE 1.16 Convert the following to vulgar fractions and simplify:

(a) $0 \cdot 6$ (b) $0 \cdot 75$ (c) $7 \cdot 5$ (d) $0 \cdot 428$ (e) $0 \cdot 002$

(a) $0 \cdot 6 \quad = \dfrac{\cancel{6}^{\,3}}{\cancel{10}_{\,5}} = \dfrac{3}{5}$

(b) $0 \cdot 75 \quad = \dfrac{\cancel{75}^{\,3}}{\cancel{100}_{\,4}} = \dfrac{3}{4}$

(c) $7 \cdot 5 \quad = 7\tfrac{5}{10} \quad = 7\tfrac{1}{2}$

(d) $0 \cdot 428 = \tfrac{428}{1000} \quad = \tfrac{107}{250}$

(e) $0 \cdot 002 = \tfrac{2}{1000} = \tfrac{1}{500}$

18

(b) Vulgar fractions to decimals

When the vulgar fractions have denominators of 10, 100, 1000, etc., the conversion is straightforward, since it is the reverse of section (a).

$$\tfrac{7}{10} = 0.7$$

$$\tfrac{43}{100} = 0.43$$

$$\tfrac{6}{100} = 0.06$$

Consider now fractions which do not contain denominators of 10, 100, etc., for example $\tfrac{1}{2}$ and $\tfrac{3}{4}$. Using the method of Section 1.7

$$\tfrac{1}{2} = \tfrac{5}{10} = 0.5$$

$$\tfrac{3}{4} = \tfrac{75}{100} = 0.75$$

Most fractions cannot be converted in this way in which case a different procedure must be used, as follows:

(a) Insert the decimal point in the numerator and add as many 0's as required.

(b) Divide the numerator into the denominator.

EXAMPLE 1.17   Convert (a) $\tfrac{3}{4}$, (b) $\tfrac{1}{8}$ to decimals.

(a) $\tfrac{3}{4}$

Write the numerator as 3.00
Divide by denominator $4\,|\,3.00$

$$\underline{0.75}$$

$$\tfrac{3}{4} = \quad 0.75$$

(b) $\tfrac{1}{8}$

Write numerator as 1.000
Divide by 8 $8\,|\,1.000$

$$\underline{0.125}$$

$$\tfrac{1}{8} = 0.125$$

From Example 1.17(b) the following conversions are worth noting:

$$\tfrac{1}{8} \qquad = 0.125$$

$$\tfrac{2}{8} = \tfrac{1}{4} = 0.25$$

$$\tfrac{3}{8} \qquad = 0.375$$

$$\tfrac{4}{8} = \tfrac{1}{2} = 0.5$$

$$\tfrac{5}{8} \qquad = 0.625$$

$$\tfrac{6}{8} = \tfrac{3}{4} = 0.75$$

$$\tfrac{7}{8} \qquad = 0.875$$

19

EXERCISE 1.13

Convert the following to vulgar fractions:

1. 0·6   2. 0·8   3. 0·09   4. 0·08
5. 0·45   6. 0·72   7. 0·33   8. 0·625
9. 0·825   10. 6·44   11. 0·01   12. 0·0001

Convert the following to decimals:

13. $\frac{3}{10}$ 14. $\frac{7}{10}$ 15. $\frac{42}{100}$ 16. $\frac{36}{100}$
17. $\frac{3}{8}$ 18. $\frac{6}{25}$ 19. $\frac{7}{20}$ 20. $\frac{5}{16}$
21. $\frac{11}{16}$ 22. $\frac{15}{16}$ 23. $\frac{7}{32}$ 24. $\frac{7}{80}$

1.14 Non-terminating decimal fractions and their reduction to a number of places

The vulgar fractions in the previous section all converted exactly into decimals. Consider, however, the two fractions

$$\tfrac{1}{13} = 0.0769\ldots$$

and

$$\tfrac{1}{7} = 0.14285\ldots$$

These two decimal fractions do not terminate at any decimal place and most vulgar fractions convert into this type of non-terminating decimal. However, it is not convenient to work with values containing a large number of decimal places so they are usually terminated after two, three, or four places. The above two numbers written with three decimal places are 0·076 and 0·142.

However, in reducing to a given decimal place, 1 must be added to it, if the first place after it is 5 or more; for example, 0·7376 reduced to two places is 0·74. The third place is 7 so that 1 must be added to the second place making 4.

Hence to three decimal places

$$\tfrac{1}{13} = 0.077, \quad \tfrac{1}{7} = 0.143$$

EXAMPLE 1.18   Reduce 12·6364 to (a) three places (b) two places of decimals.

(a) To three places, the fourth place being 4, the number becomes 12·636.
(b) To two places, the third place is 6, so 1 must be added to the second place giving the result 12·64.

A particular type of non-terminating decimal is given by

$$\tfrac{1}{3} = 0.33333...$$

and

$$\tfrac{1}{9} = 0.11111...$$

Such decimals are called **recurring** decimals.

EXERCISE 1.14

Reduce the following:

1. 7·163 to one decimal place
2. 16·763 to two decimal places
3. 21·7031 to two decimal places
4. 0·06048 to three decimal places
5. 13·7121 to no decimal places
6. 17·4653 to two decimal places
7. 108·4653 to three decimal places
8. 0·04317 to one decimal place
9. 0·05317 to one decimal place
10. 10·0603 to two decimal places

1.15 Significant figures

Instead of reducing a decimal to a number of places an alternative method is to reduce it to a given number of figures, called **significant figures**. As with decimal places, 1 is added to a figure if the next figure after it is 5 or more.

Thus

13·692 reduced to four significant figures is 13·69.

13·692 reduced to three significant figures is 13·7

Note: The number of significant figures is counted from the left.

This method applies to whole numbers as well. For example,

78263 reduced to four significant figures is 78260

78263 reduced to three significant figures is 78300

When the number is less than 1, for example 0·00127, any zeros between the decimal point and the first figure do not count towards the number of significant figures. This particular number is correct to three significant figures. Again 0·00407 is correct to three significant figures; 6·00407 on the other hand is correct to six significant figures.

EXAMPLE 1.19 Reduce the following numbers to three significant figures:

(a) 362·417 (b) 13·541 (c) 0·001706.

(a) 362·000

(b) 13·5

(c) 0·00171

EXERCISE 1.15

1. Write down the following numbers correct to three significant figures:

$$71\ 724, \quad 46\cdot551, \quad 391\cdot6, \quad 1\cdot0742, \quad 0\cdot006245$$

2. Write down the following numbers correct to two significant figures:

$$29\cdot624, \quad 1\cdot305, \quad 1\cdot091, \quad 0\cdot172, \quad 9\cdot05, \quad 0\cdot0714$$

1.16 Basic operations with decimals

(a) Addition and subtraction

This is the same as for whole numbers. It is important, however, to line up the decimal points as shown in Example 1.20.

EXAMPLE 1.20   Add 0·0612, 104·3, 10·71, 1110, 0·0003.

```
      0 · 0612
    104 · 3
     10 · 71
   1110 ·
      0 · 0003
   _____
   1225 · 0715
   _____
```

(b) Multiplication

Again this is carried out exactly as with whole numbers ignoring first of all the decimal points. The decimal point is placed in the answer in such a way that the number of decimal places is the same as the total number of decimal places in the numbers being multiplied.

EXAMPLE 1.21   Find 7·61 × 12·3

```
      7 6 1
    × 1 2 3
    _____
    2 2 8 3    × 3
    1 5 2 2    × 2
    7 6 1      × 1
    _____
    9 3 6 0 3
    _____
```

The two numbers being multiplied contain three decimal places between

22

them. Therefore the answer will contain three decimal places, that is

$$93 \cdot 603$$

(c) Division

The division of decimals is the same as for whole numbers. If the dividing number is a decimal number it must be made a whole number, by multiplying it, and the number being divided, by an appropriate number of 10's. Example 1.22 shows the procedure.

EXAMPLE 1.22   Determine

(a) $0 \cdot 0581 \div 0 \cdot 07$
(b) $68 \cdot 418 \div 12 \cdot 6$

(a) The dividing number is $0 \cdot 07$. It is made a whole number by multiplying by 100; $0 \cdot 0581$ must also be multiplied by 100. The division becomes

$$5 \cdot 81 \div 7 = 7 \underline{\smash{\big)} 5 \cdot 81}$$
$$\underline{0 \cdot 83}$$

(b) Multiplying both numbers by 10 to make $12 \cdot 6$ a whole number the division becomes

$$684 \cdot 18 \div 126 = 126 \overline{\smash{\big)} 684 \cdot 18} \quad \begin{array}{r} 5 \cdot 43 \end{array}$$

```
                    5 · 4 3
684·18 ÷ 126 = 126 | 6 8 4 · 1 8
                    6 3 0
                    ─────
                    5 4   1
                    5 0   4
                    ─────
                      3   7 8
                      3   7 8
                    ─────
                      ·   · ·
                    ─────
```

EXERCISE 1.16

Add the following:
1. $16 \cdot 34$, $0 \cdot 036$, $410 \cdot 3$
2. $100 \cdot 001$, $3 \cdot 6$, 41, $0 \cdot 0682$
3. $98 \cdot 77$, 4006, 19, $7 \cdot 082$
4. $0 \cdot 0712$, $0 \cdot 00712$, $0 \cdot 712$
5. $0 \cdot 1$, $0 \cdot 101$, $0 \cdot 0010101$, $0 \cdot 0101$

Subtract the following:

6. $111{\cdot}71 - 98{\cdot}17$
7. $0{\cdot}0571 - 0{\cdot}0399$
8. $0{\cdot}612 - 0{\cdot}0315$
9. $10 - 0{\cdot}717$
10. $1 - 0{\cdot}0604$

Evaluate the following, giving the results correct to two places of decimals:

11. $1{\cdot}12 \times 3{\cdot}14$
12. $101{\cdot}9 \times 2{\cdot}01$
13. $1{\cdot}009 \times 7{\cdot}016$
14. $171{\cdot}3 \div 4{\cdot}1$
15. $200{\cdot}76 \div 23{\cdot}01$
16. $13{\cdot}61 \times 0{\cdot}0072$
17. $4{\cdot}07 \div 0{\cdot}913$
18. $0{\cdot}797 \div 0{\cdot}0503$

Assessment test 1

1. List I shows four calculations. List II are the answers. Match the correct answers to the calculations by filling in the appropriate numbers in the boxes.

List I	List II
A. $5 + 10 \div 5 - 1 \times 2$	1. 7
B. $(5 + 10) \div 5 - 1 \times 2$	2. 5
C. $5 + 10 \div (5 - 1) \times 2$	3. 10
D. $5 + (10 \div 5 - 1) \times 2$	4. 1

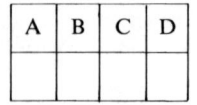

A	B	C	D

2. The laws of arithmetic are labelled as follows:

Commutative law	X
Distributive law	Y
Associative law	Z

24

State which law each of the following obeys by filling in X, Y or Z in the box.

(a) $3(4+6) = 3 \times 4 + 3 \times 6$

(b) $8 \times 9 = 9 \times 8$

(c) $2 \times (8 \times 4) = (2 \times 8) \times 4$

(d) $7+(5+6) = (7+5)+6$

3. The prime factors of 180 are $2 \times * \times 3 \times 3 \times *$. Fill in the missing numbers.

4. Calculate the LCM and HCF of 3, 9, 12.

5. The following are prime numbers; state true or false.

   (a) 6   (b) 10   (c) 17   (d) 7   (e) 12   (f) 29

6. What is the least number that 18 must be multiplied by to make it a perfect square?

   (a) 3   (b) 2   (c) 8   (d) 5

7. What is the least number that 36 must be multiplied by to make a perfect cube?

   (a) 2   (b) 3   (c) 6   (d) 7

8. The diagram shows four shaded areas, labelled A, B, C, D. The list gives four fractions corresponding to these areas. Match the fractions to the correct areas by filling in the numbers in the appropriate boxes.

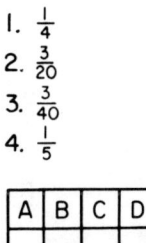

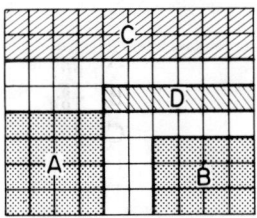

Fig. AT 1.1

9. Write down the answers to the following:

$$36 \cdot 01 \times 10$$
$$0 \cdot 0076 \times 1000$$
$$7 \times 100$$
$$0 \cdot 9 \times 1000$$

10. Evaluate the following:

$$36 \cdot 84 \div 1000$$
$$0 \cdot 0964 \div 10$$
$$0 \cdot 841 \div 100$$
$$8 \cdot 03 \div 1000$$

11. Place the fractions in order of size, starting with the biggest.

$$\tfrac{2}{3}, \quad \tfrac{5}{6}, \quad \tfrac{11}{12}, \quad \tfrac{3}{4}$$

12. Fill in the correct numbers in place of the asterisks.

(a) $\dfrac{3}{4} - \dfrac{*}{3} = \dfrac{1}{12}$

(b) $\dfrac{7}{10} \times \dfrac{*}{14} = \dfrac{3}{4}$

(c) $2\dfrac{*}{15} = \dfrac{37}{15}$

(d) $\dfrac{4}{10} \div \dfrac{7}{10} = \dfrac{4}{10} \times \dfrac{*}{*} = \dfrac{*}{*}$

(e) $\dfrac{1}{2} + \dfrac{1}{3} + \dfrac{1}{6} = *$

13. The vulgar fractions in List I are converted into decimal fractions in List II. Match the correct pairs by filling in the appropriate numbers in the boxes.

List I	List II
A. $\frac{1}{40}$	1. $0 \cdot 8$
B. $\frac{3}{100}$	2. $0 \cdot 025$
C. $\frac{5}{8}$	3. $0 \cdot 03$
D. $\frac{4}{5}$	4. $0 \cdot 625$

A	B	C	D

14. Work out

$$79{\cdot}3 \times 0{\cdot}01$$
$$8{\cdot}21 \times 0{\cdot}1$$
$$0{\cdot}0917 \div 0{\cdot}0001$$
$$13{\cdot}6 \div 0{\cdot}1$$

15. The decimal fractions in List I are converted into vulgar fractions in List II. Match the correct pairs by filling in the appropriate numbers in the boxes.

	List I		List II
A.	0·44	1.	$\frac{11}{25}$
B.	0·45	2.	$\frac{12}{25}$
C.	0·42	3.	$\frac{9}{20}$
D.	0·48	4.	$\frac{21}{50}$

A	B	C	D

16. The decimal fractions in List I are written as hundredths in List II. Match the correct pairs by filling in the appropriate numbers in the boxes.

	List I		List II
A.	0·02	1.	20
B.	0·2	2.	46
C.	0·46	3.	60
D.	0·60	4.	2

A	B	C	D

17. The number 327·652 is

.... hundreds, 2.............., ... units, 6..............,

.... hundredths, 2...............

Fill in the blanks.

18. The fraction $\frac{1}{5}$ is equal to

    (a) $10 \div 5$

    (b) $0{\cdot}2$

    (c) $5 \div 1$

    (d) $1 \times \frac{2}{10}$

    Select the **two** correct answers.

19. Since

$$\tfrac{3}{5} + \tfrac{1}{5} = \tfrac{4}{5}$$

    is

$$\tfrac{5}{3} + \tfrac{5}{1} = \tfrac{5}{4} \; ?$$

    Answer **yes** or **no**.

20. Complete the following statements:

    (a) $0{\cdot}0071$ is correct to .... decimal places

    (b) $1{\cdot}0071$ is correct to .... significant figures

    (c) $6{\cdot}41$ is correct to two ....

    (d) $18{\cdot}48$ is correct to four ....

21. State whether the following statements are true or false.

    (a) When adding mixed fractions, add the proper fractions and take the LCM of the whole numbers.

    (b) When dividing fractions change the $\div$ to $\times$ and invert the fraction on the right.

    (c) In arithmetic operations involving fractions $\times$ and $\div$ take precedence over $+$ and $-$.

    (d) When mixed fractions are multiplied, the whole numbers are multiplied first.

2. Arithmetic 2

Objectives

After working through this chapter you should be able to

1. Reduce ratios to their simplest terms.
2. Determine a quantity knowing its ratio to another quantity.
3. Calculate values of constituent parts knowing their ratios.
4. Recognize direct and inverse proportions.
5. Calculate the value of a number which is directly proportional to another number.
6. Calculate the value of a number which is inversely proportional to another number.
7. Convert fractions to percentages.
8. Convert percentages to fractions.
9. Calculate percentages.
10. Calculate actual values of given percentages.
11. Define the terms, base, index, power, reciprocal, square root, in terms of numbers with indices.
12. Multiply indexed numbers having the same base.
13. Divide indexed numbers having the same base.
14. Evaluate the power of a power of a number.
15. Relate a negative index to the reciprocal.
16. Write down the value of a number to the power 0.
17. Determine square root and cube root of numbers in index form.
18. Express decimal numbers in standard form.
19. Convert numbers in standard form to ordinary decimals.
20. Add and subtract numbers in standard form.
21. Convert an integer into binary form.
22. Convert a binary number into ordinary (denary) form.
23. Add two binary numbers.
24. Relate the binary digits to on/off switches.

2.1 Ratio

The ratio of one number 5 to another number 15 is 5/15, and is written 5 : 15. The fraction is brought down to its simplest form, so that the actual ratio is 1 : 3.

(a) To find the ratio of two quantities

Consider two firms A and B producing machine tools. Firm A turns out 30 per hour and B turns out 12 per hour. The firms produce tools in the ratio 30 : 12. The ratio in its simplest terms is 5 : 2.

EXAMPLE 2.1 Two test-tubes contain 14 g and 21 g of liquid respectively. What is the ratio of the two masses?

$$\text{Ratio of masses} = 14 : 21$$
$$= 2 : 3$$

It is important to remember that ratios of physical quantities, such as mass, length, etc., can only be determined if the values are expressed in the same units. We cannot find the ratio of two lengths 100 cm and 25 mm, without first bringing both lengths to the same unit, either both cm or both mm. Using both in cm,

$$\text{Ratio of lengths} = 100 : 2{\cdot}5$$
$$= 40 : 1, \quad \text{on dividing by } 2{\cdot}5$$

(b) Determination of quantities knowing the ratios

Type 1, where an unknown value can be found if its ratio to a known value is given. For instance, the ratio of masses of waste metal removed from two lathes A and B is 2 : 5. If the mass removed from B is 3 kg the mass removed from A can be calculated. Let the mass removed from A be m.

$$\text{Ratio of masses} = \frac{m}{3}$$

But this is known to be $\frac{2}{5}$. Therefore

$$\frac{m}{3} = \frac{2}{5}$$

Multiplying both sides by 3

$$m = \tfrac{2}{5} \times 3$$
$$= 1{\cdot}2 \text{ kg}$$

The use of ratio in this way is seen in conjunction with maps. On every map is quoted the ratio of the distance between two points on the map to the actual distance between the two points on the ground. The ratio is called the **representative fraction** (RF).

EXAMPLE 2.2   On an ordnance survey map the RF is 1 : 250 000. What is the actual distance between two towns which are 1·2 cm apart on the map?

Let the actual distance apart be s, which must be in the same units as the 1·2, that is cm. Therefore the ratio is

$$\frac{s}{1·2} = \frac{250\ 000}{1}$$

Multiplying both sides by 1·2

$$s = 250\ 000 \times 1·2$$

$$= 300\ 000\ \text{cm}$$

$$= 3\ \text{km}$$

Type 2, where a quantity is divided into parts in a given ratio. Consider a rod 180 cm long, which has to be cut in such a way that the ratio of the pieces is 4 : 5. What are the lengths of the two pieces? It is seen therefore that in this type of problem we are concerned with items which have constituent parts.

In this particular example let the rod be marked into 9 (i.e., 4+5) parts of equal length. The rod must then be cut so that the two pieces have 4 and 5 units of length, making fractions of the whole length of $\frac{4}{9}$ and $\frac{5}{9}$. Thus, the two pieces will have lengths

$$\frac{4}{9} \text{ of } 180 \quad \text{and} \quad \frac{5}{9} \text{ of } 180$$

that is,

$$80\ \text{cm} \quad \text{and} \quad 100\ \text{cm}$$

EXAMPLE 2.3   An alloy contains a mixture of three metals in the ratio 2 : 3 : 5. Calculate the mass of each metal in a quantity of 70 kg of alloy.

Because of the ratio 2 : 3 : 5 we can think of the mass of alloy as composed of 2+3+5 parts, i.e. ten parts. Thus A makes up two parts out of 10, B, three parts out of 10, C five parts out of 10.

Therefore, the quantity of A present is $\frac{2}{10}$ of 70 kg = 14 kg
the quantity of B present is $\frac{3}{10}$ of 70 kg = 21 kg
the quantity of C present is $\frac{5}{10}$ of 70 kg = 35 kg

EXERCISE 2.1

1. Find the ratios of the following quantities in simplest terms.
   (a) 6 m, 4 m   (b) 0·8 m, 1·0 m   (c) 3·2 kg, 4·8 kg   (d) 8 A, 12 A, 16 A
   (e) 12 N, 18 N, 15 N   (f) $4\frac{1}{2}$ °C, $3\frac{3}{4}$ °C   (g) 13 kg 750 g, 8 kg 250 g
   (h) 1·5 W, 0·24 W

2. Two boxes contain powder, their masses being in the ratio 3 : 7. If the second box contains 140 g, what mass is in the first?

3. The power rating of electric light bulbs in two rooms is in the ratio 4 : 9. If the total wattage in the first room is 80 W what must be the total wattage in the second room?

4. A rod is 4 m long. It is to be sawn into three pieces in the ratio 1 : 4 : 3. What are the lengths of the three pieces going to be?

5. The sides of a right-angled triangle are in the ratio 3 : 4 : 5. If the total length of the three sides is 36 cm what are the lengths of the three sides?

6. An apprentice attends 12 weeks a year at a technical college, has 2 weeks holidays and spends the remainder of the year at his work. In what ratio are these three parts of the year?

7. The volume of two cylinders are 350 mm³ and 400 mm³. Calculate the ratio of the two volumes.

8. An alloy contains three metals, X, Y and Z, in the ratio 1 : 4 : 6. Find the actual amount of each metal in a quantity of 99 kg.

9. A mixture for making concrete contains cement, sand, and stones in the ratio 2 : 5 : 7. If the quantity of sand in a given mixture is 20 kg, calculate the quantity of cement and stones in the mixture.

10. Write the following ratios in their simplest form:
   (a) 3 : 9 : 12
   (b) 14 : 21 : 35

2.2 Proportion

(a) Direct proportion

If a quantity A is directly proportional to another quantity B, then any increase in B causes an increase in A in the same ratio. If B is doubled then A is doubled. For example, the number of gallons of petrol used in a car will be directly proportional to the number of miles travelled. Consider the following figures. A motorist uses 12 gallons to travel 300 miles. How many gallons will he use to travel 200 miles?

The ratio of the number of gallons is equal to the ratio of miles, in the **direct** sense. This means that the numbers 12 and 300 appear in these ratios in the same position, either both in the numerator or both in the denominator. Let x be the number of gallons required to travel 200 miles. Therefore

$$\frac{x}{12} = \frac{200}{300}$$

Multiply both sides by 12

$$x = \frac{200}{300} \times 12 = 8 \text{ gallons.}$$

EXAMPLE 2.4   During a machining process the amount of metal removed in 8 min is 0·03 kg. How much will be removed in 12 min?

The ratio of masses removed is equal to the ratio of times, in the direct sense. Therefore, 8 min and 0·03 kg are in corresponding positions in the ratios.

Let m be the mass removed in 12 min. Thus

$$\frac{m}{0\cdot 03} = \frac{12}{8}$$

Multiply both sides by 0·03.

$$m = \frac{12}{8} \times 0\cdot 03$$

$$= 0\cdot 045 \text{ kg}$$

(b) Inverse proportion

If a quantity A is inversely proportional to another quantity B, it means that an increase in A causes a decrease in B in the same ratio.

If B is doubled then A is halved.

The electric current in a circuit decreases as the resistance increases, that is, the current is inversely proportional to the resistance. Consider the following example.

EXAMPLE 2.5   The current in a circuit is 6 Amps when the resistance is 10 ohm. What is the current when the resistance is 15 ohm?

The ratio of currents is equal to the ratio of resistances in the **inverse** sense. This means that the numbers 6 and 10 appear in these ratios in opposite positions, one in the numerator, the other in the denominator.

Let I be the current when the resistance is 15 ohm. Therefore

$$\frac{I}{6} = \frac{10}{15}$$

Multiply both sides by 6

$$I = \frac{10}{15} \times 6$$

$$= 4 \text{ A}$$

EXAMPLE 2.6   A train travels between two towns. When the speed of the train is 60 km/h it takes 4 h to complete the journey. If the speed increases to 80 km/h, calculate the new time taken to complete the journey.

The ratio of speeds is equal to the ratio of times, in the inverse sense
Therefore 60 and 4 are in opposite positions in the ratios.

Let t be the time to complete the journey at 80 km/h. Therefore

$$\frac{t}{4} = \frac{60}{80}$$

Multiply both sides by 4.

$$t = \frac{60}{80} \times 4$$

$$= 3 \text{ h}$$

EXERCISE 2.2

1. The pressure of a gas is 600 N/m² when the volume is 7·5 m³. What is the pressure when the volume is 5·0 m³?

2. A car uses $6\frac{1}{2}$ gallons of petrol for a journey of 195 miles. Find how much petrol is required for a journey of 330 miles.

3. The rise in temperature of a lump of metal is directly proportional to the quantity of heat supplied. When 400 J of heat is supplied, the temperature rise is 30 °C. What quantity of heat is supplied when the temperature rises by 45 °C.

4. Eighty articles cost £75. How much would 100 articles cost?

5. Travelling at 60 mph a car covers 400 miles in a certain time. How far would the car travel in the same time, if its speed were 45 mph?

6. A ship is 200 m long and 30 m wide at its point of greatest width. A model is to be built whose length is to be 1 m. What will be its greatest width?

7. When a car travels at 42 mph it takes 90 min to cover a journey. What speed must the car have, to cover the same journey in 70 min?

8. A wheel with 20 teeth drives a wheel with 25 teeth. The first wheel makes 40 rev/s. How many rev/s will the second wheel make?

9. A shop floor turns out finished machine tools at the rate of 40 in 3 h. Calculate how long it would take to produce 180 tools.

10. In an electrical circuit the current is proportional to the voltage. When the voltage is 3·5 V the current is 2·1 A. Calculate the current when the voltage is 6·4 V.

11. A contractor estimates that he can complete a site of houses in 484 working days employing 60 men. Calculate the increase in staff necessary to be able to complete the project in 363 working days.

2.3 Percentages

When a number is written as a percentage it means the number of parts out of 100 parts. The symbol for percentage is %. For example, 60% means 60 parts out of 100 parts, and is illustrated in Fig. 2.1 by the shaded area.

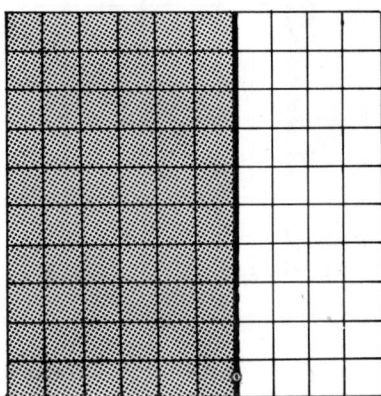

Fig. 2.1

(a) To convert % to fractions

As shown in Fig. 2.1 60% is the shaded area representing 60 parts from 100 parts. It is a fraction of the big square. Following Section 1.7, 60 parts from 100 parts is the fraction $\frac{60}{100}$. Therefore

$$60\% = \tfrac{60}{100} = \tfrac{3}{5}$$

Similarly

$$25\% = \tfrac{25}{100} = \tfrac{1}{4}$$

Therefore, to convert % to fractions divide by 100, and reduce to simplest terms.

In the same way a percentage can be converted to a decimal fraction quite easily. Again divide by 100, from which the decimal can be written down immediately.

$$50\% = \frac{50}{100} = 0.5$$

$$12.5\% = \frac{12.5}{100} = 0.125$$

35

(b) To convert fractions to %

In Fig. 2.2 the fraction $\frac{1}{5}$ is shown as the shaded area, which contains 20 squares out of 100 squares. By our definition this is 20%. Therefore

$$\tfrac{1}{5} = 20\%$$

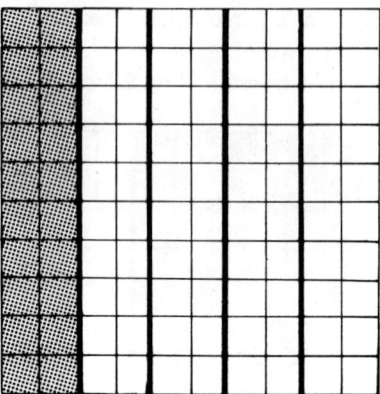

Fig. 2.2

It is seen that to convert a fraction to % multiply by 100. Hence

$$\tfrac{3}{4} = \quad \tfrac{3}{4} \times 100 = 75\%$$

$$\tfrac{1}{20} = \quad \tfrac{1}{20} \times 100 = \quad 5\%$$

$$0{\cdot}41 = 0{\cdot}41 \times 100 = 41\%$$

EXAMPLE 2.7 A substance contains the chemicals X, Y, and Z in the ratio 6 : 5 : 9. Express these as percentages.

X, Y, and Z are in the ratio 6 : 5 : 9. We can take the substance as composed of $6+5+9$, i.e., 20 parts. Therefore, fraction of X $= \tfrac{6}{20}$; fraction of Y $= \tfrac{5}{20}$; fraction of Z $= \tfrac{9}{20}$.

Therefore

$$\text{the percentage of X present} = \tfrac{6}{20} \times 100 = 30\%$$

$$\text{the percentage of Y present} = \tfrac{5}{20} \times 100 = 25\%$$

$$\text{the percentage of Z present} = \tfrac{9}{20} \times 100 = 45\%$$

EXERCISE 2.3

Convert the following to percentages:

1. $\tfrac{1}{10}$ 2. $\tfrac{3}{25}$ 3. $\tfrac{5}{8}$ 4. $\tfrac{19}{25}$

5. $\tfrac{1}{3}$ 6. $\tfrac{5}{80}$ 7. $\tfrac{3}{8}$ 8. $\tfrac{5}{12}$

9. $0{\cdot}31$ 10. $0{\cdot}52$ 11. $0{\cdot}67$ 12. $0{\cdot}05$

13. $0{\cdot}005$ 14. $0{\cdot}115$

Convert the following to fractions, both decimal and vulgar:

15. 30%   16. 45%   17. $66\frac{2}{3}$%   18. 90%
19. 80%   20. 6%   21. 0·5%   22. 0·0125%

(c) To determine percentage values

A typical problem is as follows. Thirty worn ball bearings are removed from a machine containing 75 in all. What is the percentage removed?

$$\text{Actual number removed} = 30$$

$$\text{Fraction of total removed} = \tfrac{30}{75}$$

$$\text{Percentage removed} = \tfrac{30}{75} \times 100 = 40\%$$

EXAMPLE 2.8   A man buys a car for £800 and sells it later for £700. Find his percentage loss.

$$\text{Actual loss} = £800 - £700 = £100$$

$$\text{Fractional loss} = \tfrac{100}{800}$$

Note: Profit and loss must always be expressed in terms of the buying price.

$$\text{Percentage loss} = \tfrac{100}{800} \times 100 = 12\tfrac{1}{2}\%$$

An important use of percentages in engineering and science is seen in the determination of errors in measurement.

We can think of countless examples where measurements are made in the workshop, e.g., the diameter of a hole drilled in a plate. If the actual diameter is different from the required diameter then we say that an error exists in the diameter.

(a) absolute error is the difference between the actual diameter and required diameter;
(b) fractional error is (absolute error)/(required value);
(c) percentage error is (fractional error) × 100.

Again, in experimental work in the laboratory, it is often required to calculate a quantity by making measurements. For example, the acceleration due to gravity is determined using a simple pendulum. The length and period of swing are measured. Any error in measurement will produce an error in the final result.

EXAMPLE 2.9   In the determination of the acceleration due to gravity the value obtained was 10·0 m/s². It is known that an accurate value is 9·8 m/s². Calculate the absolute, fractional, and percentage errors.

37

$$\text{Absolute error} = 10{\cdot}0 - 9{\cdot}8 = 0{\cdot}2 \text{ m/s}^2$$

$$\text{Fractional error} = \frac{0{\cdot}2}{9{\cdot}8} = \frac{1}{49}$$

$$\text{Percentage error} = \frac{1}{49} \times 100 = 2\%$$

(d) To determine actual values from given percentages

In this type of calculation the percentage of a given total has to be determined. For example, what is 2% of 3000. Such a calculation may occur in the production of 3000 transistors, where 2% are defective.

$$\text{Number of defectives} = 2\% \text{ of } 3000$$
$$= \tfrac{2}{100} \text{ of } 3000, \quad \text{where 2\% is written as a fraction}$$
$$= \tfrac{2}{100} \times 3000$$
$$= 60$$

EXERCISE 2.4

1. A man travels 120 km out of a total of 480 km. What percentage of the total distance has he covered?

2. An automated line in a firm produces 4200 pistons daily. On the average, $2\frac{1}{2}\%$ are defective. Find the total number of pistons rejected daily.

3. A drilling machine cost £120 to purchase. If, on selling, it makes 15% profit, calculate the selling price.

4. A second-hand car is bought for £320 and sold for £360. Calculate the percentage profit on the deal.

5. An ammeter has a scale with a range 0 to 100 mA. What are the limiting readings for the following currents, the maximum percentage error for each being shown in brackets:

   (a) 80 mA (1·2% of the actual current),
   (b) 15 mA (0·6% of the full-scale reading)?

   (NCTEC)

6. A typical bronze consists of 88% copper, 10% tin, and 2% zinc. Calculate the mass of each metal in 32 kg of this bronze.

   (WJEC)

7. Two machines produce similar components. One machine produced 4000 components of which 6% were rejected as unsatisfactory. The second machine produced 6000 of which 3% were rejected. Calculate the percentage of rejected components in the total output.

(WJEC)

8. A carbon resistor of nominal resistance 22 400 ohm is found to have an actual resistance of 23 800 ohm. Calculate the percentage variation from the nominal value of the resistance.

9. The resistance of a coil is marked as 14·41 ohm. When a student measures the resistance in the laboratory he obtains a result of 15·52 ohm. Calculate the actual, fractional, and percentage errors.

2.4 Indices

Consider the multiplication 10×10. This product can be written as 10^2. Following on from this, $10 \times 10 \times 10 = 10^3$. We can build a table in this way:

$$10^1 = 10$$
$$10^2 = 10 \times 10 = 100$$
$$10^3 = 10 \times 10 \times 10 = 1000$$
$$10^4 = 10 \times 10 \times 10 \times 10 = 10\ 000$$

The small number above the 10 is called the **index** and states how many times 10 must be multiplied by itself. The 10 itself is called the **base**.

(a) Multiplication of numbers to the base 10

The product $100 \times 1000 = 100\ 000$. Written with indices the product becomes $10^2 \times 10^3 = 10^5$. In order to obtain an index of 5 on the right, the two indices on the left are added together.

Rule 1. *If two numbers have the same base, multiplication is carried out by adding the indices.*

Therefore,
$$10^4 \times 10^2 = 10^6$$
$$10^3 \times 10 = 10^4$$

(b) Division of numbers of the base 10

If we divide 1 000 000 by 100, the result is 10 000. Written in index form, $10^6 \div 10^2 = 10^4$. In order to obtain the index 4 on the right the two indices on the left are subtracted.

Rule 2. *If two numbers have the same base, division is carried out by subtracting the indices*

Therefore,
$$10^7 \div 10^4 = 10^3$$
$$10^6 \div 10 = 10^5$$

(c) Multiplication and division of numbers to any base

Rules 1 and 2 do not apply to the base of 10 only. They can apply to any base number. For example,

$$2 \times 2 \times 2 = 2^3$$
$$5^4 \times 5^3 = 5^7$$
$$7^4 \div 7^3 = 7$$

(d) The power of 0

We wish to determine 7^0. Applying Rule 2

$$7^3 \div 7^3 = 7^{3-3} = 7^0$$

But, dividing a number by itself gives an answer of 1. Hence,

Rule 3. *Any number to the power 0 is equal to 1.*

(e) Negative indices

So far all the indices have been positive, but it is also possible to have a base number with a negative index. The meaning of a negative index is illustrated in the following examples

$$10^{-2} = \frac{1}{10^2}, \quad 9^{-3} = \frac{1}{9^3}$$

These results can be verified using Rules 2 and 3.

Consider $\frac{1}{9^3}$.

$$\frac{1}{9^3} = \frac{9^0}{9^3} = 9^{0-3} = 9^{-3}$$

These numbers with negative indices are called **reciprocals**.
Since

$$9^{-3} = \frac{1}{9^3}$$

9^{-3} is called the reciprocal of 9^3.

It is seen that the index changes sign when the number is taken from the top line to the bottom. The same thing happens when the number is taken from the bottom line to the top, *or when a fraction is inverted.* Consequently, we have

$$\frac{1}{10^{-2}} = 10^2, \quad \left(\frac{3}{4}\right)^{-5} = \left(\frac{4}{3}\right)^5$$

(f) Power of a power

Such numbers are of the type $(7^2)^3$. The index 3 on the outside signifies that the number inside the bracket, 7^2, must be multiplied by itself three times,

$$(7^2)^3 = 7^2 \times 7^2 \times 7^2$$
$$= 7^{2+2+2} \qquad \text{on applying Rule 1}$$
$$= 7^{3\times 2}$$
$$= 7^6$$

It is seen that such numbers are evaluated by multiplying the two powers together.

Rule 4. *A number in the form of a power of a power is evaluated by multiplying the two powers together.*

$$(6^5)^3 = 6^{5\times 3} = 6^{15}$$

(g) Square roots and cube roots

A number such as 9 can be written as

$$9 = 3 \times 3$$
$$= 3^2 \qquad \text{in index form.}$$

From Section 1.4 $\qquad \sqrt{9} = 3.$
This means that the power of 3 is halved. Therefore

$$\sqrt{9} = (3^2)^{\frac{1}{2}}$$

in which case $\qquad \sqrt{9} = 9^{\frac{1}{2}}$

Therefore, in a square root the index is $\frac{1}{2}$, that is $\sqrt{64} = (64)^{\frac{1}{2}}$. Similarly with cube roots the index becomes $\frac{1}{3}$. For example,

$$125 = 5 \times 5 \times 5$$

$$= 5^3 \qquad \text{in index form}$$

Therefore $\qquad \sqrt[3]{125} = 5 = (5^3)^{\frac{1}{3}}$

$$\sqrt[3]{125} = (125)^{\frac{1}{3}}$$

EXAMPLE 2.10 Find the following roots by writing the numbers, first of all, in index form $\qquad \sqrt{64}, \quad \sqrt[3]{216}, \quad \sqrt{\frac{4}{9}}$

$$\sqrt{64} = \sqrt{2^6} = (2^6)^{\frac{1}{2}} = 2^3 = 8$$

$$\sqrt[3]{216} = \sqrt[3]{6^3} = (6^3)^{\frac{1}{3}} = 6^1 = 6$$

$$\sqrt{\frac{4}{9}} = \sqrt{\frac{2^2}{3^2}} = \left(\frac{2^2}{3^2}\right)^{\frac{1}{2}} = \frac{2}{3}$$

EXERCISE 2.5
Using the two rules of indices simplify the following, leaving the result in index form:

1. $10^2 \times 10^3$

2. $4^6 \times 4^2$

3. $3^9 \times 3$

4. $9^2 \times 9 \times 9^3$

5. $4^3 \div 4$

6. $9^4 \div 9^2$

7. $\dfrac{4^2 \times 4^3}{4 \times 4^2}$

8. $\dfrac{8^2 \times 6^8}{8 \times 6^7}$

9. $\dfrac{7^6 \times 7^3 \times 4^8}{7^8 \times 4^2 \times 4^3}$

Write out the following with positive indices, leaving the result in index form:

10. 2^{-2}

11. 10^{-3}

12. 8^{-1}

13. $\dfrac{1}{2^{-4}}$

14. $\dfrac{1}{10^{-3}}$

15. $\dfrac{1}{9^{-1}}$

Evaluate the following:

16. 2^0 17. 6^0 18. $(18)^0$ 19. 4×7^0

20. 6×9^0 21. $\dfrac{1}{10^0}$ 22. $(4^6 - 3^7)^0$ 23. $(19^{10} - 10^{19})^0$

Using the rule for a power of a power, evaluate the following, leaving the result in index form:

24. $(2^2)^3$ 25. $(3^4)^2$ 26. $(10^3)^5$ 27. $(9^3)^7$

28. $(6^7)^2$ 29. $(6^7)^2 \times (6^4)^3$ 30. $(2^3)^5 \times (2^2)^4$

Evaluate the following:

31. $\sqrt{81}$ 32. $\sqrt{64}$ 33. $\sqrt{\frac{1}{16}}$ 34. $\left(\frac{9}{64}\right)^{\frac{1}{2}}$

35. $\left(\frac{25}{49}\right)^{\frac{1}{2}}$ 36. $\left(\frac{9}{25}\right)^{-\frac{1}{2}}$ 37. $\left(\frac{4}{9}\right)^{-\frac{1}{2}}$ 38. $\sqrt[3]{27}$

39. $\sqrt[3]{\frac{1}{64}}$ 40. $\left(\frac{27}{64}\right)^{\frac{1}{3}}$ 41. $\left(\frac{8}{125}\right)^{\frac{1}{3}}$ 42. $\left(\frac{64}{125}\right)^{-\frac{1}{3}}$

2.5 Numbers expressed in standard form

Decimal numbers may be expressed in the following manner:

$$12 \cdot 1 = 1 \cdot 21 \ \times 10 \ = 1 \cdot 21 \ \times 10$$
$$227 \cdot 6 = 2 \cdot 276 \times 100 \ = 2 \cdot 276 \times 10^2$$
$$1462 = 1 \cdot 462 \times 1000 = 1 \cdot 462 \times 10^3$$

The numbers in the last column are written in standard form.
When the numbers are less than one, we have the standard forms as follows:

$$0 \cdot 621 = \frac{6 \cdot 21}{10} = \frac{6 \cdot 21}{10^1} = 6 \cdot 21 \times 10^{-1}$$

$$0 \cdot 00174 = \frac{1 \cdot 74}{1000} = \frac{1 \cdot 74}{10^3} = 1 \cdot 74 \times 10^{-3}$$

EXAMPLE 2.11   Find, without using tables, the value of

$$\frac{4 \cdot 5 \times 10^{-4} \times 1 \cdot 2 \times 10^8}{9}$$

giving the answer in standard form.
First of all we bring the 10^{-4} to the bottom, i.e.,

$$\frac{4 \cdot 5 \times 1 \cdot 2 \times 10^8}{9 \times 10^4}$$

43

The tens are cancelled out by subtracting the indices, i.e.,

$$\frac{5 \cdot 4 \times 10^4}{9} = 6 \cdot 0 \times 10^3$$

EXERCISE 2.6

Express the following numbers in standard form:

1. $361 \cdot 1$, 4216, $33 \cdot 62$

2. $0 \cdot 00367$, $0 \cdot 217$, $0 \cdot 0417$

3. Evaluate without using tables:

(a) $\dfrac{7 \cdot 5 \times 10^2 \times 1 \cdot 2 \times 10^3}{4}$

(b) $\dfrac{3 \cdot 3 \times 10^{-4} \times 4 \cdot 2 \times 10^8}{1 \cdot 1 \times 10^{-3} \times 2 \cdot 1 \times 10^2}$

4. Express the following numbers in normal decimal form.

(a) $3 \cdot 612 \times 10^2$ (b) $4 \cdot 316 \times 10^3$ (c) $2 \cdot 6 \times 10^4$

(d) $5 \cdot 81 \times 10^{-2}$ (e) $6 \cdot 217 \times 10^{-1}$ (f) $8 \cdot 22 \times 10^{-5}$

Addition and subtraction of numbers in standard form

Two numbers in standard form can only be added or subtracted if the power of 10 is the same in both cases. For example:

$$6 \cdot 14 \times 10^2 + 2 \cdot 1 \times 10^2 = 8 \cdot 24 \times 10^2$$
$$2 \cdot 17 \times 10^{-3} - 1 \cdot 92 \times 10^{-3} = 0 \cdot 25 \times 10^{-3}$$

If the powers of 10 are not the same they must first be made equal before addition or subtraction can be carried out. Addition cannot be carried out with the two numbers

$$6 \cdot 14 \times 10^3 + 2 \cdot 1 \times 10^2$$

until both powers of 10 have been made the same, that is,

$$6 \cdot 14 \times 10^3 + 0 \cdot 21 \times 10^3 = 6 \cdot 35 \times 10^3$$

EXERCISE 2.7

Add and subtract the following numbers in standard form.

1. $3 \cdot 16 \times 10^3 + 2 \cdot 135 \times 10^3$

2. $4 \cdot 145 \times 10^{-2} - 3 \cdot 02 \times 10^{-2}$

3. $7 \cdot 63 \times 10^{-4} + 4 \cdot 45 \times 10^{-4}$

4. $6 \cdot 01 \times 10^2 - 4 \cdot 10 \times 10^2$

5. $7 \cdot 9 \times 10^2 + 1 \cdot 05 \times 10^2 - 3 \cdot 005 \times 10^2$

6. $6 \cdot 65 \times 10^2 + 3 \cdot 098 \times 10^3$

7. $4 \cdot 55 \times 10^{-3} + 1 \cdot 23 \times 10^{-2}$

8. $6 \cdot 85 \times 10^{-2} - 7 \cdot 53 \times 10^{-3}$

9. $7 \cdot 012 \times 10^{-4} - 1 \cdot 2 \times 10^{-5}$

10. $5 \cdot 0 \times 10^2 + 5 \cdot 0 \times 10^{-2}$

2.6 Binary numbers

In our normal counting system, called the **denary** system, there are 10 characters,

$$0, 1, 2, \ldots, 9$$

Such a system is said to have a base of 10. Consider a number such as 429. It can be written as

$$429 = \text{four sets of } 100 + \text{two sets of } 10 + \text{nine units}$$
$$= 4 \times 100 + 2 \times 10 + 9 \times 1$$
$$= 4 \times 10^2 + 2 \times 10^1 + 9 \times 10^0$$

It is seen that each digit, moving from left to right, signifies 10^0, 10^1, 10^2, etc.

The binary system has only two characters, 0 and 1. It is said to have a base of 2. Numbers have the same meaning as in denary, but now signify 2^0, 2^1, 2^2, etc., that is, 2 replaces the 10. Therefore the binary number 1111 can be written as

$$1 \times 2^3 + 1 \times 2^2 + 1 \times 2^1 + 1 \times 2^0$$

Similarly 1001 is

$$1 \times 2^3 + 0 \times 2^2 + 0 \times 2^1 + 1 \times 2^0$$

(a) Binary counting

When 1 is added to 9 in denary, the sum reverts back to the digits 10, that is $9 + 1 = 10$, where 9 is the highest character.

In binary the highest character is 1. Adding 1 to it makes the sum revert back to the digits 10, that is

$$1 + 1 = 10$$

Counting in any system is really the addition of 1 to the last number all the time. This is shown in the following tables, where the carry-over figure is shown below the line.

The denary equivalent is shown below each sum

	0	1	10	11	100	101	110	111	1000
	1	1	1	1	1	1	1	1	1
	—	—	—	—	—	—	—	—	—
Binary	1	10	11	100	101	110	111	1000	1001
	—	—	—	—	—	—	—	—	—
		1				1		11	carry over

| Denary | 1 | 2 | 3 | 4 | 5 | 6 | 7 | 8 | 9 |

(b) Conversion of binary to denary

First the number is written in powers of 2 as already explained. Then all that is required is to work out these powers.

For example,

$$1111 = 1 \times 2^3 + 1 \times 2^2 + 1 \times 2^1 + 1 \times 2^0$$
$$= 8 \quad + \quad 4 \quad + \quad 2 \quad + \quad 1$$
$$= 15$$

$$1001 = 1 \times 2^3 + 0 \times 2^2 + 0 \times 2^1 + 1 \times 2^0$$
$$= 8 \quad + \quad 0 \quad + \quad 0 \quad + \quad 1$$
$$= 9$$

(c) Conversion of denary to binary

Each digit moving from left to right in binary means 1, 2, 4, 8, 16, 32, 64. Thus any denary number must be expressed as a sum of all or some of the above numbers. For example,

$$59 = 32 + 27$$
$$= 32 + 16 + 11$$
$$= 32 + 16 + 8 + 3$$
$$= 32 + 16 + 8 + 4 + 2 + 1$$
$$= 1 \quad 1 \quad 1 \quad 0 \quad 1 \quad 1$$

Note that the sum of 59 is made up without the 4, shown as 4. Each number represents a 1 in binary, except 4, which represents 0, as shown below the line. Therefore 59 in binary is

$$111011$$

EXAMPLE 2.12   Convert 35 into binary number.

$$35 = 32 + 3$$
$$= 32 + \cancel{16} + \cancel{8} + \cancel{4} + 2 + 1$$
$$\overline{= 1 \quad 0 \quad 0 \quad 0 \quad 1 \quad 1}$$

16, 8 and 4 are not required to make up the number. Therefore the binary number is 100011.

(d) Addition in binary

Addition is merely an extension of counting in binary, using the result that

$$1 + 1 = 10$$

The 1 in the 10 will appear as a carry-over figure, and is shown in Section 2.6(a).

EXAMPLE 2.13   Add 10101 and 11101

```
    1 0 1 0 1
    1 1 1 0 1
  _____
  1 1 0 0 1 0
  _____
    1 1 1   1      carry over
```

EXERCISE 2.8

1. Convert the following to denary numbers:
   (a) 110111   (b) 101   (c) 10011   (d) 100001
   (e) 10000   (f) 11010   (g) 100101   (h) 11011

2. Convert the following to binary numbers:
   (a) 11   (b) 24   (c) 10   (d) 29
   (e) 19   (f) 64   (g) 32   (h) 45

3. Carry out addition of the following binary numbers:
   (a) 101   (b) 101   (c) 1110
        10   11   1011
       ___   ___   ____

   (d) 10111   (e) 110111   (f) 11111
        1011   10110   1
       _____   _____   ____

47

(e) The use of the binary system

The binary system is so strange to us that it is important to ask what practical use is there to it. In modern technology the binary method of counting is connected very largely with computers.

A computer can only deal with bits of information which is a straight choice, that is where the answer is **yes** or **no, true** or **false**. The bits of information are said to be **two-state** systems. Any problem must be broken down into a sequence of two-state steps.

The reason for this is that the computer circuits can only be two-state systems. A circuit can only be closed, when a current will flow, or open, when no current will flow.

The arithmetic required to match this property of the circuitry must also be two-state, that is, it must be a binary system, with only two types of digits in it, 0 and 1. Fig. 2.3 shows a switch in two states, (a) being closed corresponding to the digit 1, and (b), being open corresponding to 0.

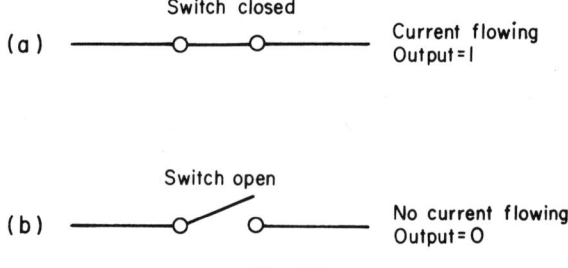

Fig. 2.3

Assessment test 2

1. Two numbers are in the ratio 18 : 12. Reduce the ratio to simplest terms.

2. Two rods have lengths 2 m 50 cm and 6 m 25 cm. Choose the simplest form of their ratio from the following:

    (a) 2 : 7

    (b) 50 : 25

    (c) 2 : 1

    (d) 2 : 5

3. Ratios are given in List I. In List II these are reduced to their simplest form. Match List II with List I by filling in the appropriate numbers in the boxes.

List I	List II
A. 15 : 20 : 25	1. 4 : 5 : 3
B. 20 : 25 : 15	2. 2 : 1 : 3
C. 8 : 16 : 12	3. 3 : 4 : 5
D. 9 : 4½ : 13½	4. 2 : 4 : 3

A	B	C	D

4. Two rods are shown in the diagram. If B is 14 cm long what is the length of A?

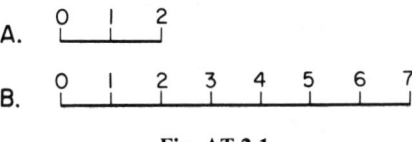

Fig. AT 2.1

5. The two areas are in the ratio 4 : 9. If the shaded area is 8 m² what is the area of the unshaded area?

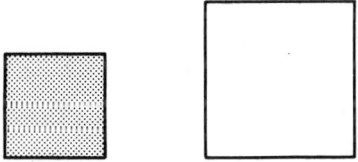

Fig. AT 2.2

6. A certain food contains protein, fat, carbohydrate, in the ratio 1 : 2 : 5. What is the actual weight of each in 24 kg of the food.?

7. (a) What is the square root of 9^4 in index form?
   (b) What is the cube root of 5^6 in index form?
   (c) Evaluate $(\frac{49}{100})^{\frac{1}{2}}$.

8. When the temperature of a rod increases by 20 °C the length increases by 0·4 mm. If the rod increases in length by 0·5 mm what was the temperature increase?

9. The electric current I in a circuit varies inversely as the resistance R. If i and I are the values of the current at resistances of 10 ohm and 20 ohm respectively, what is the ratio i/I equal to?

10. List I shows the same area shaded in different fractions. List II gives the same shaded areas as percentages. Match List II with List I by filling in the appropriate numbers in the boxes.

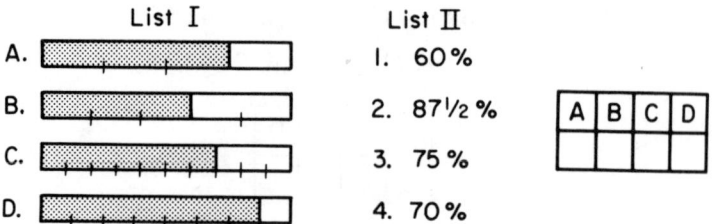

Fig. AT 2.3

11. Convert (a) 25%, (b) 24% to vulgar fractions.

12. Convert (a) 14%, (b) 6·7% to decimal fractions.

13. Twenty-one out of a class of 24 students were successful in the TEC first year examination. Which of the following gives the correct percentage?

(a) 21%

(b) $87\frac{1}{2}$%

(c) $12\frac{1}{2}$%

(d) $24\frac{1}{2}$%

14. The diameter of a wheel is 200 mm. It is measured as 204 mm. What is the percentage error?

15. Three per cent of items produced in a factory are defective. In a batch of 4000 find the number that will be rejected when tested.

16. In the number 7^3 state the value of base, index, power.

17. For the number 8^2 state the reciprocal and the square root?

18. In the following complete the calculation by putting the correct number in the square.

(a) $6^5 \times 6^4 = 6^{\square}$

   $7^3 \times 7^{\square} = 7^8$

(b) $8^8 \div 8 = 8^{\square}$

   $9^{10} \div 9^{\square} = 9^2$

(c) $(8^3)^5 = 8^{\Box}$

        $(3^{\Box})^6 = 3^{24}$

(d) $8^{\Box} = 1$

        $(6^0)^5 = \Box$

19. List I shows numbers in normal decimal form. List II gives the powers of 10 when these numbers are expressed in standard form. Match List II to List I by filling in the numbers in the boxes.

	List I		List II
A.	161·3	1.	1
B.	0·00712	2.	2
C.	91·6	3.	-1
D.	0·3412	4.	-3

A	B	C	D

20. Evaluate the following

(a) $3 \times 10^3 + 6 \times 10^3 - 2 \times 10^3$

(b) $4 \times 10^2 + 7 \times 10^3 + 4 \times 10^{-1}$

21. The number 57 can be written as $32 + 16 + 8 + 0 + 0 + 1$. Which of the following is the correct binary form?

(a) 1 1 1 1 1 1

(b) 1 1 1 0 0 1

(c) 1 1 1 0 1 0

(d) 1 1 1 1 0 0

3. Calculations

Objectives

After working through this chapter you should be able to

1. Explain why exact values can be obtained when counting but not when measuring.
2. State the possible error in any number expressed to a given number of significant figures.
3. Reject an answer to a calculation if it is not feasible.
4. Give the answer to a calculation to a reasonable number of significant figures.
5. State why a 'guarding' figure is used during calculations.
6. Check the answer to a calculation by determining its approximate value.
7. List the types of aids available for numerical calculations.
8. Determine square roots, squares, and reciprocals of four-figure numbers using four-figure tables.
9. Define the logarithm to the base 10.
10. Deduce that the logarithm of unity is zero.
11. Use four-figure logarithm tables to
    (a) determine the logarithm of a positive number
    (b) determine the antilogarithm of a logarithm
    (c) multiply and divide numbers
    (d) evaluate roots and reciprocals of numbers
    (e) evaluate the value of a number raised to an integral power.
12. Identify the parts of a slide rule.
13. Use a slide rule to:
    (a) multiply and divide numbers
    (b) determine square roots, squares, and reciprocals of numbers.
14. Use an electronic calculator to
    (a) add, subtract, multiply, and divide numbers
    (b) determine the value of a number raised to an integral power
    (c) determine the reciprocal of a number.

3.1 Introduction

Engineers and scientists must carry out many calculations as part of their work. For example, it might be necessary to calculate the size of a bolt to carry a certain weight.

It is obviously very important to carry out calculations without making arithmetic mistakes. It must be emphasized that such mistakes can be avoided if the work is written down neatly and methodically.

An answer obtained from a calculation should not be accepted without first checking that it is a feasible solution. For example, if the length of a bolt were calculated to be 20 m, it is obvious that such a solution would not be feasible. It suggests that either

(a) an arithmetic mistake had been made in the calculation or

(b) the original information was incorrect.

On the other hand, even if the answer seems feasible it is still good practice to check back through the calculation.

Calculations can become long and tedious, and various aids are available to reduce the time spent on them. These aids are discussed later in the chapter, and it is necessary to be aware of their advantages and disadvantages.

3.2 Errors and accuracy

When counting it is possible to obtain an exact answer, such as, for example, counting the number of bolts in a box. If the box contains 62 bolts it is possible by counting carefully to obtain an exact answer of 62.

With measured quantities, however, such as length, time, weight, etc., it is impossible to know if, and indeed unlikely that, an exact value has been obtained. The accuracy of the answer depends upon the accuracy of the measuring instrument, such as a ruler or a clock, and also upon the eyesight of the person reading the instrument.

If the length of a rod is measured as 14·6 cm it does not follow that its length is exactly equal to 14·6 cm. It simply means that the length of the rod is nearer to 14·6 cm than to either 14·5 cm or 14·7 cm. Since three figures are used in the measurement 14·6 cm, the length is said to be measured to an accuracy of three significant figures. Any length between 14·55 cm and 14·65 cm would be measured as 14·6 cm, correct to three significant figures. Another way of looking at it is, that the value 14·6 cm, correct to three significant figures, could have an error up to 0·05 cm on either side of this value.

If this value 14·6 cm is now used in a calculation, then the error in the value will produce an error in the answer to the calculation.

Consider the calculation to find the area of a rectangular-shaped floor, of length 15·7 m and width 4·3 m.

$$\text{Area} = \text{length} \times \text{width}$$
$$= 15\cdot7 \times 4\cdot3$$
$$= 67\cdot51 \ \text{m}^2$$

This answer is correct only if the length is exactly 15·7 m and the width is exactly 4·3 m. But

the length 15·7, correct to three significant figures, may have any value between 15·65 m and 15·75 m

and

the width 4·3 m, correct to two significant figures, may have any value between 4·25 m and 4·35 m.

Using the pair of lower values and the pair of higher values, the area may have any value between

$$15·65 \times 4·25 \quad \text{and} \quad 15·75 \times 4·35$$

that is, between 66·5125 m² and 68·5125 m².

It is seen that due to possible errors in the measurement of length and width the error in the area could be in the second significant figure. It would be misleading to give the answer for the area as 67·51 m², since this suggests that it is accurate to four significant figures.

To avoid such misleading answers the following rule is adopted.

Rule. *The number of significant figures in an answer is limited to one* **more** *than the least number of significant figures used in the original information.*

This rule would apply to the four arithmetic operations, $+, -, \times, \div$.

In the above calculation of area, the least accurate information was two significant figures in the width. Therefore, the answer is given correct to three significant figures, that is,

$$\text{area} = 67·5 \text{ m}^2$$

The extra significant figure (that is 5) in the answer is called a **guarding** figure. It is needed if the answer is used in a further calculation, otherwise the second calculation would produce an answer accurate only to the one significant figure.

3.3 Approximate values

When an answer has been obtained it is good practice to check quickly for arithmetic mistakes, and to find out if the answer is at least approximately correct.

An approximate answer can be quickly obtained by 'rounding off' all numbers to fewer significant figures, as shown in Example 3.1.

EXAMPLE 3.1   Find the approximate value of

$$\frac{7·2 \times 9·8}{3·7 \times 2·14}$$

Rounding off the numbers gives,

$$\text{approximate value} = \frac{7 \times 10}{4 \times 2} = \frac{35}{4} \backsimeq 9$$

The exact value of the answer is 8·91, showing that good agreement is possible. If the answers do not agree the calculation must be re-checked. One important application of finding an approximate value is that it fixes the position of the decimal point.

3.4 Aids to numerical calculations

The following aids are available to the engineer and scientist to reduce the work involved in numerical calculations.

(a) Mathematical tables

These include logarithm and antilogarithm tables, square root tables, reciprocal tables, etc. Four-figure tables are normally used, that is, tables with an accuracy of four significant figures. Tables with higher degrees of accuracy are available for specialist work.

(b) Slide rule

(c) Electronic calculator

3.5 Evaluation of square roots, squares, and reciprocals by tables

Each of these may be obtained by normal arithmetic methods but they can be obtained much more easily using the appropriate tables.

Square roots

It is known that

$$\begin{aligned} \sqrt{10} &= 3 \cdot 162 \\ \sqrt{100} &= 10 \end{aligned} \Big\}$$
$$\begin{aligned} \sqrt{1000} &= 31 \cdot 62 \\ \sqrt{10\,000} &= 100 \end{aligned} \Big\}$$

By comparing the above square roots it can be seen that if the decimal point is moved one place the numbers in the square root are completely different but if the decimal point is moved two places the numbers in the square root are the same. For this reason the square root tables are in two parts, namely 1–10 and 10–100.

To obtain the square root of a number the decimal point must be moved *two* places at a time until the number can be read from the tables. The decimal point must then be moved back one place in the answer for every two places it was originally moved. This process can be easily done by boxing the numbers in pairs from the point and placing one figure of the answer in each box.

EXAMPLE 3.2   Find the value of (a) $\sqrt{40}$ and (b) $\sqrt{400}$.

(a) This can be read directly from the tables:

$$\sqrt{40} = 6\cdot325$$

(b) Box in pairs from the point $\sqrt{4{,}00}\cdot$
   Read $\sqrt{4\cdot00} = 2\cdot00$ from tables and
   place one figure in each box $2{\mid}0$:
   Therefore, $\sqrt{400} = 20$.

EXAMPLE 3.3   Evaluate $\sqrt{12\ 76}$
   Box in pairs from the point $\sqrt{12{\mid}76}\cdot$
   Read $\sqrt{12\cdot76} = 3\cdot572$ from tables
   and place one figure in each box $3{\mid}\ 5{\mid}72$
   Therefore $\sqrt{1276} = 35\cdot72$

Approximate value $35 \times 35 = 1225$, that is nearly 1276.
The method of obtaining $\sqrt{12\cdot76}$ is shown in the reprint of part of the square root tables (Fig. 3.1).

SQUARE ROOTS. From 10 to 100

	0	1	2	3	4	5	6	7	8	9	Mean Differences		
											123	456	7 8 9
10	3.161	3.178	3.194	3.209	3.225	3.240	3.256	3.271	3.286	3.302	235	689	11 12 14
11	3.317	3.332	3.347	3.362	3.376	3.391	3.406	3.421	3.435	3.450	134	679	10 12 13
12	3.464	3.479	3.493	3.507	3.521	3.536	3.550	3.564	3.578	3.592	134	678	10 11 13
13	3.606	3.619	3.633	3.647	3.661	3.674	3.688	3.701	3.715	3.728	134	578	10 11 12
14	3.742	3.755	3.768	3.782	3.795	3.808	3.821	3.834	3.847	3.860	134	578	9 11 12
15	3.873	3.886	3.899	3.912	3.924	3.937	3.950	3.962	3.975	3.987	134	568	9 10 11

Fig. 3.1

$$\sqrt{12\cdot76} = 3\cdot564$$

                    8 (Mean Difference)   Add
                    ―――――
                    $3\cdot572$

EXAMPLE 3.4   Evaluate $\sqrt{0{\mid}00{\mid}73{\mid}12}$
Box in pairs from the point
Read $\sqrt{73\cdot12} = 8\cdot551$ from tables
and place one figure in each box, $\cdot\ 0{\mid}\ 8{\mid}\ 5{\mid}51$
i.e., $\sqrt{0\cdot007312} = 0\cdot08551$.
Approximate value $0\cdot08 \times 0\cdot08 = 0\cdot0064$.

Squares. Since squaring a number is the reverse process of finding the square root then squares can be found by using the square root tables.

$$\text{Also, since} \quad 4^2 = 16$$

$$\text{and} \quad 40^2 = 1600$$

it is seen that the actual numbers in the answer are independent of the position of the point. The decimal point can therefore be ignored when using the tables and its position in the answer can easily be determined by finding its approximate value.

EXAMPLE 3.5   Evaluate $76 \cdot 74^2$.

The method is shown in Fig. 3.2

SQUARE ROOTS. From 10 to 100

	0	1	2	3	4	5	6	7	8	9	Mean Differences		
											123	456	7 8 9
55	7.416	7.423	7.430	7.436	7.443	7.450	7.457	7.463	7.470	7.477	112	334	5 5 6
56	7.483	7.490	7.497	7.503	7.510	7.517	7.523	7.530	7.537	7.543	112	334	5 5 6
57	7.550	7.556	5.563	7.560	7.576	7.583	7.589	7.596	7.603	7.609	112	334	5 5 6
58	7.616	7.622	7.629	7.635	7.642	7.649	7.655	7.662	7.668	7.675	112	334	5 5 6
59	7.681	7.688	7.694	7.701	7.707	7.714	7.720	7.727	7.733	7.740	112	334	4 5 6
60	7.746	7.752	7.759	7.765	7.772	7.778	7.785	7.791	7.797	7.804	112	334	4 5 6

Fig. 3.2

$$7668^2 = \quad 588$$
$$6 \qquad\quad 9$$
$$\overline{} \quad \overline{}$$
$$7674^2 \quad 5889$$

The decimal point has to be placed in the answer.
Approximate value $80^2 = 6400$.
Therefore $76 \cdot 74^2 = 5889$.
Square tables are also available and may be used in the normal way to find squares of numbers.

Reciprocals. The reciprocal of a number is the value of 1 divided by the number, i.e., the reciprocal of 2 is $\frac{1}{2}$ which is $0 \cdot 5$.

Reciprocal tables are used in the same way as the other tables except that the Mean Difference must be *subtracted*. This is because the greater the number the smaller will be its reciprocal. The position of the decimal point does not affect the actual figures in the answer but the reciprocal table gives the reciprocals of standard numbers so it is preferable to arrange the numbers in standard form.

EXAMPLE 3.6   Find the reciprocal of 5·493.

This can be found directly from the tables

$$
\begin{array}{ll}
5\cdot49 & 0\cdot1821 \\
\quad\; 3 & \qquad 1 \;\text{(note subtraction)} \\
\hline
5\cdot493 & 0\cdot1820
\end{array}
$$

Therefore $1/5\cdot493 = 0\cdot1820$.
Approximate value $\frac{1}{5} = 0\cdot2$.

EXAMPLE 3.7   Find the value of $1/672\cdot4$.

Arranging in standard form, $\dfrac{1}{672\cdot4} = \dfrac{1}{6\cdot724} \times \dfrac{1}{10^2}$

From tables $\dfrac{1}{6\cdot724} = 0\cdot1487$

Therefore $\dfrac{1}{672\cdot4} = 0\cdot1487 \times \dfrac{1}{10^2}$

$= 0\cdot001487$

EXAMPLE 3.8   Find the value of $1/0\cdot04824$.

In standard form $\dfrac{1}{0\cdot04824} = \dfrac{1}{4\cdot824} \times \dfrac{1}{10^{-2}}$

From tables $\dfrac{1}{4\cdot824} = 0\cdot2073$

Therefore $\dfrac{1}{0\cdot04824} = \dfrac{0\cdot2073}{10^{-2}}$

$= 20\cdot73$

Approximate value $1/0\cdot05 = 20$.

EXERCISE 3.1

Use the appropriate tables to do this exercise. Use approximate values to check your answers.

1. Find the square roots of:
   2·279, 836·2, 25 000, 0·337, 0·0529

2. Find the squares of the following numbers, expressing your answers in standard form:
   1·115,   26·45,   0·1833,   0·0927

3. Find the reciprocals of:
   2·8,   341·4,   0·9835,   0·006339

4. Evaluate:

   (a) $\sqrt{0·00443}$ (b) $(0·3636)^2$ (c) $\dfrac{1}{0·0621}$

5. Evaluate:

   (a) $\sqrt{(2·29^2 + 6·224^2)}$ (b) $\dfrac{1}{3·15^2}$

   (c) $\dfrac{1}{\sqrt{0·0252}}$ (d) $\dfrac{1}{4·152} + \dfrac{1}{3·147}$

   (e) $\sqrt{\left(\dfrac{1}{0·6323} + \dfrac{1}{0·3625}\right)}$

3.6 Logarithms

It has been seen in Chapter 2 that when numbers with the same base are multiplied or divided the indices are added or subtracted. These facts can be used to simplify complicated calculations by expressing all numbers as powers of a common base. Tables have been compiled which enable these powers to be obtained quickly. These tables are known as **logarithmic tables** and the value obtained from the tables is known as the **logarithm** of the number. Any base may be chosen for a set of tables, and with **common** logarithms this base is 10.

The logarithm of a number to a given base is the power to which the base must be raised to equal the number

Since $100 = 10^2$, then by definition the logarithm of 100 on a base 10 is 2. This may be written $\log_{10} 100 = 2$. For normal calculations common logarithms are used and the 10 is usually omitted, i.e., $\log 100 = 2$.

Similarly $1000 = 10^3$, i.e., $\log_{10} 1000 = 3$ or $\log 1000 = 3$. Since logarithms are indices (powers) they obey the same rules:

(a) for multiplying numbers, add indices, add logs;

(b) for dividing numbers, subtract indices, subtract logs.

Consider the problem $100 \times 1000 = 100\ 000$. In index form, $10^2 \times 10^3 = 10^5$. If the problem is done by logs:

Number	Log
100	2
1000	3
Adding	5

This 5 is still a log, therefore to obtain the answer, i.e., 100 000, a reverse process to that of obtaining logs is required. **Antilogarithm tables** are used for this process and the number obtained from these tables is called an **antilogarithm**. It is the answer to the problem but without the decimal point.

To obtain the logarithms. Since log 100 $= 2$, log 1000 $= 3$, etc., any number in between 10, 100, 1000, etc., will have a decimal logarithm. A logarithm is therefore seen to consist of two parts as follows.

(a) The number before the point (called the **characteristic**) is found by inspection. The characteristic is the number of places the decimal point has to be moved to put the number into standard form.

EXAMPLE 3.9

Number	Standard form	Characteristic
49·2	$4·92 \times 10^1$	1
397·2	$3·972 \times 10^2$	2
6279	$6·279 \times 10^3$	3

Note: The characteristic is always equal to the index in the standard form.

(b) The number after the point (called the **mantissa**) is found from the logarithm tables.

Note: The decimal point is ignored when using the log tables.

EXAMPLE 3.10   Find the mantissa for the number 136·5.

LOGARITHMS

	0	1	2	3	4	5	6	7	8	9	1	2	3	4	5	6	7	8	9
10	0000	0043	0086	0128	0170						5	9	13	17	21	26	30	34	38
						0212	0253	0294	0334	0374	4	8	12	16	20	24	28	32	36
11	0414	0453	0492	0531	0569						4	8	12	16	20	23	27	31	35
						0607	0645	0682	0719	0755	4	7	11	15	18	22	26	29	33
12	0792	0828	0864	0899	0934						3	7	11	14	18	21	25	28	32
						0969	1004	1038	1072	1106	3	7	10	14	17	20	24	27	31
13	1139	1173	1206	1239	1271						3	6	10	13	16	19	23	26	29
						1303	1335	1367	1399	1430	3	7	10	13	16	19	22	25	29
14	1461	1492	1523	1553	1584						3	6	9	12	15	19	22	25	28
						1614	1644	1673	1703	1732	3	6	9	12	14	17	20	23	26
15	1761	1790	1818	1847	1875						3	6	9	11	14	17	20	23	26
						1903	1931	1959	1987	2014	3	6	8	11	14	17	19	22	25

Fig. 3.3

The mantissa is obtained from the tables as shown in Fig. 3.3.

Number	Mantissa	
136	1335	
5	16	
1365	0·1351	Add

i.e., the mantissa is 0·1351.

Since $136·5 = 1·365 \times 10^2$, i.e., the characteristic $= 2$, then

$$\log\ 136·5 = 2·1351.$$

EXAMPLE 3.11

Number	Characteristic	Mantissa	Logarithm
12·19	1	0·0860	1·0860
432·7	2	0·6362	2·6362
2·040	0	0·3096	0·3096

EXERCISE 3.2

1. Find the characteristic of the following numbers:
   21,   1000,   2·95,   10·51,   300·67

2. Find the mantissa of the following numbers:
   2250,   367·4,   8·437,   99·81,   54·246

3. Find the logarithms of the following numbers:
   19·1,   707·2,   4·439,   4629·0,   86·428

To obtain the antilogarithms. Since only the mantissa is obtained from the logarithm tables, it is only the mantissa that is used in the antilogarithm tables to obtain the antilogarithm. The characteristic indicates where the decimal point is to be placed in the final answer.

EXAMPLE 3.12   Find the antilogarithm of 2·5466.

ANTILOGARITHMS

	0	1	2	3	4	5	6	7	8	9	123	4 5 6	7 8 9
.50	3162	3170	3177	3184	3192	3199	3206	3214	3221	3228	112	3 4 4	5 6 7
51	3236	3243	3251	3258	3266	3273	3281	3289	3296	3304	122	3 4 5	5 6 7
52	3311	3319	3327	3334	3342	3350	3357	3365	3373	3381	122	3 4 5	5 6 7
53	3388	3396	3404	3412	3420	3428	3436	3443	3451	3459	122	3 4 4	6 6 7
54	3467	3475	3483	3491	3499	3508	3516	3524	3532	3540	122	3 4 5	6 6 7
.55	3548	3556	3565	3573	3581	3589	3597	3606	3614	3622	122	3 4 5	6 7 7
56	3631	3639	3648	3656	3664	3673	3681	3690	3698	3707	123	3 4 5	6 7 8

Fig. 3.4

The antilogarithm is obtained from the tables as shown in Fig. 3.4.

Mantissa	Antilogarithm	
0·546	3516	
6	5	
0·5466	3521	Add

To obtain the position of the decimal point in the final answer, reverse the process that is used for finding the characteristic. Put the antilogarithm in standard form and for a positive characteristic move the decimal point that number of places to the right.

For Example 3.12 the characteristic = 2, so start with standard form, i.e., 3·521, and move two places to the right; then the answer is 352·1.

Check the following using antilogarithm tables:

Logarithm	Antilogarithm	Number
4·3575	2278	22 780
2·9100	8128	812·8
0·3181	2080	2·080

EXERCISE 3.3

1. Find the antilogarithms of the following:
   5·1496,   0·7002,   1·0121,   2·9100,   4·3

2. Use antilogarithm tables to change the following logarithms back into ordinary numbers:

   3·1295,   2·4437,   1·0095,   0·3002,   2·0910,   0·0101

3.7 Multiplication and division using logarithms

Multiplication

Rule. *When multiplying numbers ADD the logarithms.*

EXAMPLE 3.13   Evaluate $34\cdot2 \times 12\cdot4$.

Number	Logarithm
34·2	1·5340
12·4	1·0934

$$2\cdot6274$$

Antilogarithm   4240

Therefore $34\cdot2 \times 12\cdot4 = 424\cdot0$. (Approximate value $34 \times 12 = 408$.)

Division

Rule. *When dividing numbers SUBTRACT the logarithms.*

EXAMPLE 3.14   Evaluate $\dfrac{24\cdot7}{2\cdot92}$

Number	Logarithm
24·7	1·3927
2·92	0·4654

$$0\cdot9273$$

Antilogarithm   8459

Therefore $\dfrac{24\cdot7}{2\cdot92} = 8\cdot459$. $\left(\text{Approximate value } \dfrac{24}{3} = 8.\right)$

Combined multiplication and division. With calculations of this type the steps to follow are:

(a) work out the top line in logarithmic form;
(b) work out the bottom line in logarithmic form;
(c) subtract the logarithm of the bottom line from the logarithm of the top line;
(b) find the antilogarithm.

63

EXAMPLE 3.15 Find the value of

$$\frac{347 \cdot 2 \times 18 \times 31 \cdot 4}{19 \cdot 8 \times 206}$$

	Number	Logarithm
Top line 347·2		2·5406
18		1·2553
31·4		1·4969

Logarithm of top line =   5·2928 ⟶ 5·2928

Bottom line   19·8   1·2967 ⟶ 3·6106

206   2·3139

                                              1·6822

Logarithm of bottom line =   3·6106

Antilogarithm   4810

Therefore $\dfrac{347 \cdot 2 \times 18 \times 31 \cdot 4}{19 \cdot 8 \times 206} = 48 \cdot 10.$

$$\left(\text{Approximate value } \frac{350 \times 20 \times 30}{20 \times 200} = 52.\right)$$

Note: Only the multiplication and division of numbers can be done by logarithms. Addition and subtraction of numbers is done in the normal way.

EXERCISE 3.4

The student is advised to practise the use of approximate values in this exercise, wherever feasible.

1. Use logarithms to find the value of the following products, expressing the answer in standard form:

   (a) $3 \cdot 947 \times 18 \cdot 1$ (b) $64 \cdot 7 \times 19 \cdot 5$
   (c) $200 \cdot 4 \times 637$ (d) $14 \cdot 3 \times 8600$

2. Evaluate the following, correct to three significant figures:

   (a) $\dfrac{32 \cdot 9}{12 \cdot 2}$ (b) $\dfrac{1 \cdot 119}{1 \cdot 087}$ (c) $\dfrac{31\ 920}{33 \cdot 8}$ (d) $\dfrac{886\ 952}{485}$

3. Evaluate, using logarithms:

   (a) $647 \times 285 \times 3 \cdot 142$ (b) $\dfrac{207 \cdot 2 \times 13 \cdot 4}{59 \cdot 79}$

(c) $\dfrac{663\cdot4}{2\cdot15\times266}$ (d) $\dfrac{24\cdot4\times37\cdot6}{17\cdot42\times40\cdot2}$

4. Evaluate, correct to three significant figures, using logarithms when necessary:

(a) $\dfrac{16\cdot24+3\cdot12}{2\cdot49}$ (b) $\dfrac{464}{125}-\dfrac{84\cdot5}{63\cdot2}$

(c) $\dfrac{234(662-359)}{528\cdot4}$ (d) $\dfrac{4\cdot95\times47\cdot6}{18\cdot2+19\cdot3}$

(e) $867\cdot2\div64\cdot8+9\cdot154$

3.8 Logarithms of numbers less than 1

Consider the following numbers less than 1:

Number		Index form
0·1	=	10^{-1}
0·01	=	10^{-2}

Since a logarithm is an index then

$$\log 0\cdot1 = -1$$
$$\log 0\cdot01 = -2$$

It is seen that the characteristic for a number less than 1 can be obtained in exactly the same way as for a number greater than 1 except for the minus sign. The minus sign is placed over the characteristic and is read as bar 1, bar 2, etc. Since the minus sign is over the characteristic then *only* the characteristic is negative, the mantissa is *always* positive and is determined from the tables in the normal way.

EXAMPLE 3.16 Find the logarithm of 0·0384.

The logarithm of 0·0384 equals the logarithm of $3\cdot84\times10^{-2}$, i.e., the characteristic is $\bar{2}$.

From tables, the mantissa $= 0\cdot5843$, i.e., $\log 0\cdot0384 = \bar{2}\cdot5843$.

Check the following table:

Number	Standard form	Characteristic	Mantissa	Logarithm
0·794	$7\cdot94\times10^{-1}$	$\bar{1}$	0·8998	$\bar{1}\cdot8998$
0·00295	$2\cdot95\times10^{-3}$	$\bar{3}$	0·4698	$\bar{3}\cdot4698$
0·0176	$1\cdot76\times10^{-2}$	$\bar{2}$	0·2455	$\bar{2}\cdot2455$

Note: The addition and subtraction of bar numbers must follow the algebraic laws of addition and subtraction established in Chapter 4.

If a number is carried over from the mantissa to the characteristic in the bottom line, then it is always advisable to sort out the bottom line before proceeding.

EXAMPLE 3.17   Evaluate $27 \cdot 3 \times 0 \cdot 01961$.

Number	Logarithm
$27 \cdot 3$	$1 \cdot 4362$
$0 \cdot 01961$	$\bar{2} \cdot 2925$

$$\bar{1} \cdot 7287$$

Since it is addition, simply collect the characteristics, i.e., 1 and $-2 = -1$ or $\bar{1}$.

Antilogarithm   5355

Therefore $27 \cdot 3 \times 0 \cdot 01961 = 5 \cdot 355 \times 10^{-1} = 0 \cdot 5355$.
(Approximate value $30 \times 0 \cdot 02 = 0 \cdot 6$.)

EXAMPLE 3.18   Evaluate $\dfrac{297}{0 \cdot 2102}$.

Number	Logarithm
297	$2 \cdot 4728$
$0 \cdot 2102$	$\bar{1} \cdot 3226$

$$3 \cdot 1502$$

Since it is subtraction, change the sign of the characteristic on the bottom line, i.e., -1 to $+1$, then collect, i.e., $+1$ and $2 = +3$.

Antilogarithm   1414

Therefore $\dfrac{297}{0 \cdot 2102} = 1 \cdot 414 \times 10^3 = 1414$.

$\left(\text{Approximate value } \dfrac{300}{0 \cdot 2} = 1500.\right)$

EXERCISE 3.5

1. Use logarithms to evaluate the following products, leaving your answer in standard form: Check your answers using approximate values.

   (a) $16 \cdot 3 \times 0 \cdot 2954$ (b) $0 \cdot 1334 \times 202 \cdot 4$

   (c) $42 \cdot 5 \times 0 \cdot 0567$ (b) $0 \cdot 1423 \times 0 \cdot 02914$

   (e) $0 \cdot 0064 \times 0 \cdot 00888$

2. Evaluate, using logarithms:

(a) $\dfrac{32\cdot47}{0\cdot8386}$ (b) $\dfrac{0\cdot3984}{315\cdot9}$ (c) $\dfrac{0\cdot7639}{0\cdot147}$ (d) $\dfrac{0\cdot4381}{0\cdot7964}$

(e) $\dfrac{0\cdot01487}{0\cdot0434}$

3. Use logarithm tables to evaluate the following:

(a) $\dfrac{0\cdot0359\times2\cdot612}{0\cdot0531}$ (b) $\dfrac{0\cdot005382}{0\cdot0277\times0\cdot1493}$ (c) $\dfrac{650\cdot2\times0\cdot0791}{2\cdot115\times0\cdot3324}$

4. Find the value of the following expressions, correct to three significant figures:

(a) $\dfrac{0\cdot007}{0\cdot052}+1\cdot312$ (b) $\dfrac{0\cdot1133\times15\cdot49+1\cdot279}{0\cdot04336}$

(c) $\dfrac{0\cdot1838\times0\cdot9781}{1\cdot4437-1\cdot3926}$ (d) $\dfrac{1}{0\cdot0266}+\dfrac{16\cdot4}{2\cdot33}$

3.9 Logarithm of unity

In Section 2.4 it was seen that

$$10^0 = 1$$

Since logarithms and indices are the same thing it follows that

$$\log_{10} 1 = 0$$

3.10 Reciprocals using logarithm tables

Reciprocals can be obtained by using logarithm tables and the result $\log_{10} 1 = 0$. This is shown in Example 3.19

EXAMPLE 3.19   Find the reciprocal of 18.2

It is required to find $\dfrac{1}{18\cdot2}$.

This is a division and can therefore be done using logarithm tables.

Number	Logarithm
1	0·0000
18·2	1·2601
	$\bar{2}$·7399

Antilogarithm   5494

Therefore $\dfrac{1}{18 \cdot 2} = 0 \cdot 05494.$ $\left(\text{Approximate value} = \dfrac{1}{20} = 0 \cdot 05.\right)$

EXERCISE 3.6   Use logarithm tables to find the reciprocal of:

1. 2·8   2. 341·4   3. 0·9835   4. 0·006339

3.11 Powers and roots

Powers and roots can be evaluated easily using logarithms.

Powers

EXAMPLE 3.20.   Evaluate $3 \cdot 972^2$

Now $3 \cdot 972^2 = 3 \cdot 972 \times 3 \cdot 972.$

Number	Logarithm
3·972	0·5990
3·972	0·5990
	1·1980

Antilogarithm   1578

Therefore $3 \cdot 972^2 = 15 \cdot 78.$ (Approximate value $4 \times 4 = 16.$)

Note: The answer can also be found by multiplying $\log 3 \cdot 972$ by 2. This leads to the rule for finding powers by logarithms.

Rule. *To find the value of a number raised to a power, MULTIPLY the logarithm of the number by the power and find the antilogarithm to obtain the answer.*

Roots

Since roots can be expressed as fractional powers (see Section 2.4) then the following rule applies.

Rule. *To find the root of a number, DIVIDE the logarithm of the number by the root and find the antilogarithm to obtain the answer.*

EXAMPLE 3.21   Find the cube root of 2812.

Number	Logarithm
2812	3⌷ 3·4490
	1·1497

Antilogarithm   1411

Therefore $\sqrt[3]{2812} = 14\cdot11$. (Approximate value $14^3 = 196 \times 14$ $\simeq 200 \times 14 = 2800$.)

EXERCISE 3.7

Check answers using approximate values.

1. Use logarithm tables to find the squares of the following numbers, expressing the answers in standard form:

   1·495, 18·64, 92·62, 151·5, 2041

2. Find the cubes of the following numbers, putting the answers in standard form:

   3·556, 9·276, 14·5, 52·15

3. Find the square roots of the following numbers using logarithm tables, giving your answers correct to three significant figures:

   1·325, 12·9, 160, 62 660

4. Evaluate

   (a) $1\cdot437^4$ (b) $\sqrt[5]{104\cdot7}$ (c) $1\cdot449^3 \times 13\cdot2$

   (d) $\dfrac{\sqrt[3]{85\cdot1}}{3\cdot47}$ (e) $\sqrt{(41\cdot2^2 + 23\cdot15^2)}$

3.12 Powers and roots of numbers less than 1

In Section 3.11 powers and roots of numbers greater than 1 were dealt with. This work must now be extended to deal with the powers and roots of numbers less than 1. Remember that algebraic rules must be applied when dealing with bar numbers.

Powers

EXAMPLE 3.22 Evaluate $0\cdot0937^3$.

Number	Logarithm
0·0937	$\bar{2}\cdot9717$

$3 \times -2 = -6$ and 2 carried over from mantissa gives -4, i.e., $\bar{4}$

	3

$\bar{4}\cdot9151$

Antilogarithm	8224

Answer $= 0\cdot0008224$

69

Roots. If the characteristic is negative it must be *exactly* divisible by the root. This is because a negative number cannot be carried over to the positive mantissa. If the negative characteristic is not exactly divisible by the root it must be modified until it is, by changing it to the next *highest* divisible number.

Case I: If the characteristic is exactly divisible.

EXAMPLE 3.23   Find the value of $\sqrt{0.0186}$.

$$
\begin{array}{cc}
\text{Number} & \text{Logarithm} \\
0.0186 & 2\lfloor\overline{2}\cdot2695 \\
& \overline{1}\cdot1348 \\
\text{Antilogarithm} & 1364
\end{array}
$$

Answer $= 0.1364$

(Approximate value $0.14^2 = 0.0196$.)

Case II: If the characteristic is not exactly divisible.

EXAMPLE 3.24   Find the fourth root of 0.006964.

The characteristic must be changed to $\overline{4}$ by adding $\overline{1}$. To keep the size of the characteristic the same $+1$ must also be added, i.e., $\overline{3} = \overline{4}+1$
The division now proceeds

$$
\begin{array}{cc}
\text{Number} & \text{Logarithm} \\
0.006964 & \overline{3}\cdot8428 \\
& 4\lfloor\overline{4}+1\cdot8428 \\
& \overline{1}+0\cdot4607 \\
& \overline{1}\cdot4607 \\
\text{Antilogarithm} & 2889
\end{array}
$$

Answer $= 0.2889$

(Approximate value $0.3^4 = 0.0081$.)

EXERCISE 3.8

Use logarithm tables in this exercise. Check using approximate values.

1. Evaluate:
    (a) $\sqrt{0.03784}$ (b) $\sqrt{0.1143}$
    (c) $\sqrt[3]{0.0219}$ (d) $\sqrt[3]{0.0001485}$
    (e) $\sqrt{0.00479}$ (f) $\sqrt[4]{0.0614}$

2. Find the value of:
    (a) $(0.1424)^2$ (b) $(0.4126)^3$
    (c) $(0.0214)^2$ (d) $(0.966)^5$

70

3.13   The slide rule

The slide rule is equivalent to a set of logarithm tables because the numbers marked on the scales of the rule are placed at distances proportional to the **mantissa** of their logarithms. It is therefore possible to do those calculations with the slide rule that could have been done by logarithm tables, that is, to multiply and divide two numbers, and to find the powers, roots, and reciprocals of numbers. Operating instructions are supplied with the slide rule.

There are three main parts of the slide rule (as shown in Fig. 3.5). These are the *stock*, *slider*, and *cursor* (which is a sliding marker).

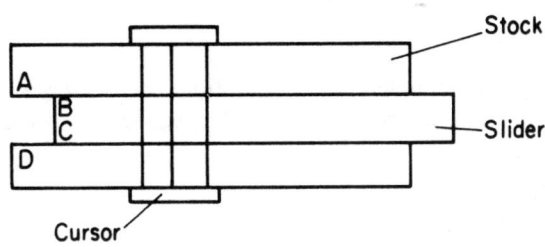

Fig. 3.5

When two numbers are multiplied together their logarithms are added together. For example, to multiply 2 by 3 the logarithms of 2 and 3 are added together. Similarly with a slide rule, to multiply two numbers the slider is moved so that a length on scale D (proportional to the log of one number) and a length on scale C (proportional to the log of the other number) are added together. The answer is obtained on scale D. This is shown in Fig. 3.6 with the multiplication 2×3. The cursor is used to match the scales on the stock and slider.

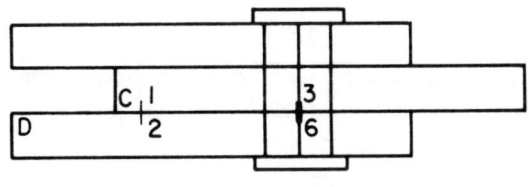

Fig. 3.6

To divide two numbers the logarithms are subtracted, so that on the slide rule lengths are subtracted on scales C and D, to obtain the answer on scale D. The division $8 \div 4$ is shown in Fig. 3.7.

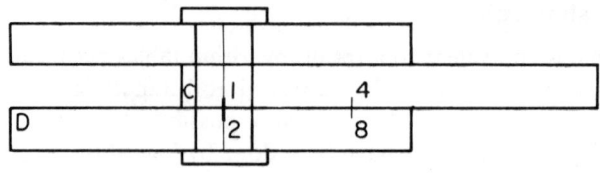

Fig. 3.7

The answers are accurate to two or three significant figures, depending upon which parts of the scales are being used. However, the slide rule does not give the position of the decimal point in the final answer. This can be found by the approximate method described in Section 3.2.

EXERCISE 3.9

Use a slide rule to complete those questions in Exercises 3.1–3.7 which involve multiplication, division, square roots, squares, and reciprocals.

3.14 Calculators

Pocket electronic calculators are now available at reasonable prices. These are capable of rapidly calculating answers to 10 or 12 significant figures. It must be remembered, however, that this level of accuracy is rarely required in engineering, since few measurements in engineering have accuracies greater than four or five significant figures. An answer to a calculation can only be as accurate as the original information.

The numbers and operations in a calculation are entered by pressing keys. All numbers entered and the final answer are displayed on the number display panel. Detailed instructions are supplied with each calculator and slight differences occur in the key sequence of different calculators. The following examples illustrate the key operations for one make of calculator.

Example	Key operation sequence
(a) $6+4$	6 + 4 =
(b) $6-4+7$	6 − 4 + 7 =
(c) $6\times4\div5$	6 × 4 ÷ 5 =
(d) $\frac{1}{6}$	1 ÷ 6 =
(e) $\sqrt{47\cdot2}$	4 7 · 2 √

In all calculations involving mixed operations the rules of preference discussed in Section 1.1 must be followed. These rules of preference are summarized by the word BODMAS, as explained in Section 1.1. Consider,

72

for example, the calculation $16 + 8 \div 4$. If the key operation sequence

is followed the incorrect answer of 6 is obtained. This comes about because the rules of precedence BODMAS are not followed.

The correct sequence is

$$\boxed{8}\ \boxed{\div}\ \boxed{4}\ \boxed{+}\ \boxed{1}\ \boxed{6}\ \boxed{=}$$

giving an answer of 18.

Polish notation. Some calculators use a sequence which is called the polish notation. The calculation $6 - 4$ will have the key operation sequence

$$\boxed{6}\ \boxed{+}\ \boxed{4}\ \boxed{-}$$

In other words, the sequence is as if numbers were added and subtracted from a total in store. The $+$ and $-$ operator is always added after the number.

EXERCISE 3.10

Use an electronic calculator to complete those questions in Exercises 3.1–3.7, which involve addition, subtraction, multiplication, division, numbers raised to integral powers (that is, a whole number), reciprocals, and if you have a square root key on your calculator, square roots.

3.15 Comparison of aids

A set of four figure logarithm tables provides a cheap aid to carrying out calculations, at the same time making them easier to do.

A slide rule is more expensive, but it is far quicker to use. The slide rule, however, is not as accurate as the four-figure tables, but for many purposes in engineering its accuracy is sufficient.

Electronic calculators provide answers accurately and quickly, and with modern prices are relatively cheap. By using batteries these calculators are portable, especially as they are pocket size. These modern calculators are certainly the most useful of the aids to calculations.

Assessment test 3

1. The length of a bar is given as 3·7 m. State the shortest possible length of the bar.
    (a) 3·6 m
    (b) 3·65 m
    (c) 3·7 m
    (d) 3·75 m

2. An engine casting has a mass of 600 kg, correct to two significant figures. The maximum possible mass of the casting is
    (a) 590 kg
    (b) 595 kg
    (c) 599 kg
    (d) 605 kg

3. A technician completes a job in 52 h, his next job in 14 h. If these times are recorded to the nearest hour what is the shortest time in which he could have completed both jobs?
    (a) 65 h
    (b) 65·5 h
    (c) 66 h
    (d) 66·5 h

4. The answer to a numerical question should usually be given to the following number of significant figures,
    (a) one
    (b) two
    (c) three
    (d) One more than the least accurate data.

5. What is $\log_{10} 1$ equal to?
    (a) 0
    (b) 0·1
    (c) 1
    (d) 10

6. Which of the following operations may be carried out using logarithms?
    (a) addition
    (b) subtraction
    (c) multiplication
    (d) division

7. If $60 = 10^{1·7782}$, what is $\log_{10} 60$?

8. The reciprocal of 2 is
   (a) 0·2
   (b) 0·5
   (c) −0·5
   (d) 2^{-1}
   Choose **two** correct answers.

9. Given that $1·7^2 = 2·89$ what is $\sqrt{2·89}$?

10. Give approximate values for the following:
    (a) $\dfrac{15·7 \times 7·2}{640 \times 50}$
    (b) $\sqrt{180}$
    (c) $\sqrt{3607}$
    (d) $(41·43)^2$

11. Complete the following statements.
    (a) When multiplying two numbers, their logarithms must be ....
    (b) The logarithm of a number to base 10 is the .... to which the base must be raised to equal that number.
    (c) The logarithm of .... is zero.
    (d) When finding the square root of a number, its logarithm must be divided by ....

12. Which aid(s) may be used to subtract two numbers?
    (a) logarithm tables   (b) slide rule   (c) calculator

13. Given that $\sqrt{4} = 2$, $\sqrt{40} = 6·325$, without using tables find the value of
    (a) $\sqrt{400}$
    (b) $\sqrt{4000}$
    (c) $\sqrt{0·4}$
    (d) $\sqrt{0·04}$

14. $\sqrt{0·016}$ is
    (a) 0·4
    (b) 0·04
    (c) 0·126
    (d) 0·0126

15. List I gives four numbers and List II the characteristics of their logarithms. Match these characteristics to the numbers by filling in the appropriate numbers in the boxes.

List I	List II
A. 3710	1. $\bar{1}$
B. 0·00371	2. 3
C. 0·371	3. $\bar{3}$
D. 3·71	4. 0

A	B	C	D

16. Which of the following statements are correct?

(a) If $\log_{10} y = \bar{1}\cdot 8451$ then $y = 0\cdot 07$.
(b) If $\log_{10} y = 2\cdot 4901$ then $y = 309\cdot 1$.
(c) If $\log_{10} y = 0\cdot 9586$ then $y = 0\cdot 9090$.
(d) If $\log_{10} y = 1\cdot 0004$ then $y = 10\cdot 01$.

17. List I contains four numbers. List II gives the number of significant figures in each. Match List II to List I by filling in the appropriate numbers in the boxes.

List I	List II
A. 0·0071	1. 4
B. 3124	2. 5
C. 5·10	3. 3
D. 10·007	4. 2

A	B	C	D

18. Use logarithms to calculate

(a) $8\cdot 72 \times 17\cdot 1$ (b) $\dfrac{8\cdot 72}{17\cdot 1}$ (c) $8\cdot 72^3$ (d) $\dfrac{1}{17\cdot 1}$ (e) $\sqrt{17\cdot 1}$

19. Use four-figure tables other than logarithms to calculate

(a) $4\cdot 6^2$ (b) $\sqrt{4\cdot 6}$ (c) $\dfrac{1}{4\cdot 6}$

20. Which of the following calculations could be carried out with the slide rule in the position shown in Fig. AT 3.1?
    (a) $1 \cdot 8 \times 6$
    (b) $1 \cdot 8 \div 6$
    (c) $3 \div 6$
    (d) 3×6

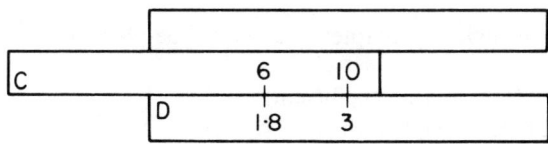

Fig. AT 3.1

4. Algebra 1—Basic operations

Objectives

After working through this chapter you should be able to

1. Recognize a term and its coefficient.
2. Add and subtract simple terms.
3. Apply the commutative, associative, and distributive laws.
4. Multiply and divide simple terms by arithmetic numbers.
5. Apply the rules of precedence.
6. Substitute numbers for letters in algebraic expressions.
7. Add, subtract, multiply, and divide directed numbers.
8. Define the terms base, index, power, reciprocal, root for algebraic numbers.
9. Multiply indexed numbers having the same base.
10. Divide indexed numbers having the same base.
11. Evaluate the power of a power of a number.
12. Relate a negative index to the reciprocal.
13. Evaluate numbers with fractional indices.
14. Apply the distributive law to expansion of brackets
    (a) where the bracket is multiplied by a number or letter
    (b) where the bracket is multiplied by another bracket.
15. Factorize expressions by
    (a) extracting a common factor.
    (b) grouping in pairs.

4.1 Letters, and their addition and subtraction

Algebra is essentially the mathematics of letters which are used to represent numbers. It is an extension of arithmetic and uses many more types of operations.

In arithmetic repeated addition of the same number can be carried out by multiplication. Thus

$$6+6+6+6 = 4 \times 6$$

Similarly in algebra we have

$$z+z+z+z = 4 \times z$$

We can give to z a value 6 or any other appropriate number. It is customary to omit the $\times$ sign in numbers like $4 \times z$, merely writing them as $4z$.

More than two numbers may be multiplied together. The product of the four numbers 5, a, b, c, is $5 \times a \times b \times c$ which is written as $5abc$.

The numbers $4z$ and $5abc$ are called **terms**. The number 4 in front of z or the 5 in front of abc is called the **coefficient**. If the expression contains more than one term, the individual terms are separated by $+$ or $-$ signs. In the expression

$$4x + 7xy - 9yz$$

there are three terms, viz. $4x$, $7xy$, $9yz$.

Consider, now, the addition of two terms such as

$$4z + 2z$$

Writing this sum out in full we obtain

$$z + z + z + z + z + z = 6z$$

The addition is carried out by adding the coefficients, that is,

$$5t + 7t = 12t$$

Similarly with the subtraction of two terms the coefficients are subtracted.

$$8a - 5a = 3a$$

In these examples so far the terms being added or subtracted contain the same letter. When the terms contain different letters addition and subtraction cannot be carried out. Thus the sum

$$z + x$$

cannot be taken further, and must, therefore, be left as it is. The rule may be stated as follows,

Rule. *Terms containing the same letters may be added and subtracted. Terms containing different letters cannot be added or subtracted.*

The application of this rule is shown in the following examples.

EXAMPLE 4.1   Evaluate the following:

(a) $9y + 4y - 7y$
(b) $4s + 16t + 5s - 3t$

   (a) $9y + 4y - 7y = 13y - 7y$ Step 1: Add the first two terms
                   $= 6y$ Step 2: Subtract the resulting two terms
   (b) The addition of the symbols s is carried out, i.e.

$$4s + 5s = 9s$$

The subtraction of the symbols t is carried out, i.e.,

$$16t - 3t = 13t$$

so that the result is

$$9s + 13t$$

EXERCISE 4.1

In the following examples carry out the addition or subtraction wherever possible:

1. $4x + 3x$
2. $8p - 6p$
3. $4x + 7y$
4. $8t - 9s$
5. $5t + 2t + 7t$
6. $5c + 4c - 7c$
7. $7e - 3e + 6e$
8. $9y + 3y + 5a - 2a$
9. $3t + 7m + 5t - 4m$
10. $5a + 3a - 6a + 3b - 2b + 7b$
11. $7p + 8p + 5q + 2p + 2q - 5p$
12. $9z - 8z$
13. $14t - 14t$

4.2 Three laws of algebra

The three laws are identical to the three laws of arithmetic listed in Section 1.2.

(a) The commutative law

This law states that in the addition or multiplication of two numbers the order in which they are written down does not affect the answer. Therefore

$$a \times b = b \times a$$
$$3 \times z = z \times 3$$
$$x + y = y + x$$
$$4 + t = t + 4$$

This law has already been used in Example 4.2(b). In this worked example the $16t + 5s$ in effect change positions,

$$4s + 5s + 16t - 3t = 9s + 13t$$

(b) The associative law

The law states that in the addition or multiplication of three or more numbers, the order in which the operation is carried out does not affect the answer. Therefore

$$x+(y+z) = (x+y)+z$$

$$u \times (v \times w) = (u \times v) \times w$$

As a consequence of the commutative law it is customary to write the product of two or more algebraic numbers in the order that they occur in the alphabet. For example, $8xday$ is more convenient to use when it is written as $8adxy$.

EXAMPLE 4.2   Simplify the following:

$$5pmn + 7npm - 2mpn$$

Using the commutative law the terms are re-written with the letters in the order of the alphabet.

$$5mnp + 7mnp - 2mnp = 10mnp$$

EXERCISE 4.2

Simplify the following, wherever possible:

1. $5ab+3ab$
2. $12vxz-9vxz$
3. $5xy+7uz$
4. $5am \quad 12apq$
5. $8ks+7kt$
6. $5xy+7yx$
7. $13ut-12tu$
8. $17vrt-13tvr$
9. $8axy+4xay$
10. $8xy+7mn-6xy+4mn$
11. $13ab+19uv-10ab-7uv$
12. $13af+12xy-3xy-7af$
13. $9gi+10ft+4ft-9gi$

(c) The distributive law

Extending the results of Section 1.2(c) to letters, we may write

$$a(x+y) = a \times x + a \times y = ax+ay$$

$$3(s+t) = 3 \times s + 3 \times t = 3s+3t$$

$$2(a-b) = 2 \times a - 2 \times b = 2a-2b$$

These three results illustrate the distributive law.

81

EXERCISE 4.3

Use the distributive law to expand the following:

1. $5(p+q)$
2. $7(u+v-n)$
3. $t(1+m+n)$
4. $5(-r-s+t)$
5. $2a(y-z)$

6. $r(2a+2b-6c)$
7. $mn(pq-rs)$
8. $4x(y+z)$
9. $2y(-x-z)$
10. $5z(ab-bc)$

4.3 Simple multiplication and division

As discussed already an addition of numbers such as $3+3+3+3$ may be performed far more easily when expressed in the form 4×3. Similarly, $3r+3r+3r+3r$ may be expressed as $4 \times 3r$ to give $12r$. In this type of operation the coefficient of r is multiplied by 4.

EXAMPLE 4.3   Work out each of the following:

(a) $7t+7t+7t$
(b) $5 \times 8v$
(c) $1 \times 11m$
(d) $0 \times 12e$

As stated above:

(a) $7t+7t+7t = 3 \times 7t = 21t$
(b) $5 \times 8v = 40v$
(c) $1 \times 11m = 11m$
(d) $0 \times 12e = 0$, remembering that multiplying any number by 0 gives an answer of 0.

Division is carried out in a similar manner. If $14f \div 7$ is required, the 14 is divided by 7, the result being $2f$.

EXERCISE 4.4

Simplify the following:

1. $5x \times 7$
2. $18t \div 9$
3. $yx \times 9$
4. $3m \times 0$
5. $21t \div 7$

6. $3\frac{1}{2}z \times 2$
7. $12uvw \times 12$
8. $44xz \div 11$
9. $12uvw \div 12$
10. $cdf \times 10$

4.4 The rules of precedence

As in arithmetic when a mixture of operations occur in algebraic calculations the rules of precedence determine the order in which the operations are carried out.

Rule 1. $\times$ *and* $\div$ *are carried out before* $+$ *and* $-$.

For example, in the expression

$$4 \times z + 3 \times z$$

the multiplication is carried out first to give

$$4z + 3z = 7z$$

Again,

$$3 \times 2t + 5 \times 2t - 4 \times 3t = 6t + 10t - 12t$$
$$= 4t$$

Rule 2. *Whenever brackets occur in a calculation, the operations inside the brackets are carried out first.*

In the calculation

$$4(2z + 5z)$$

the sum inside the bracket is worked out first, to give

$$4 \times 7z = 28z$$

EXERCISE 4.5

Evaluate the following expressions using the rules of precedence:

1. $3m \times 3 + 2m \times 5$
2. $8p \div 4 + 12p \div 6$
3. $7 \times 5p - 8 \times 3p$
4. $3 \times 4f + 2 \times 6f - 5 \times 2f$
5. $\frac{3}{2} \times 4f + \frac{3}{4} \times 20f - \frac{3}{10} \times 30f$
6. $9t \times 2 - 12t \div 6$
7. $5(5a + 6a)$
8. $r(6s + 2s)$
9. $m(3n + 2n \times 4)$
10. $3(18p \div 9 - 2p)$

4.5 Simple substitution

In Section 4.2 it was emphasized that letters in algebra represent numbers. If, in any expression, each letter is given a particular value the process is called **substitution**. It is illustrated in Example 4.4.

EXAMPLE 4.4   Calculate the value of $6abc - 4a + 3c$ when $a = 2$, $b = 3$, and $c = 5$.

On making the substitution

$$6abc - 4a + 3c = 6 \times 2 \times 3 \times 5 - 4 \times 2 + 3 \times 5$$
$$= 180 - 8 + 15$$
$$= 187$$

EXERCISE 4.6

Calculate the value of each of the following expressions, when $m = 4$, $s = 2$, $r = 3$:

1. $m + s + r$
2. $2m + 8s - 5r$
3. $3m - 2s + 3r$
4. $2m - 4s + 3r$
5. $5mrs$

6. $ms + sr - mr$
7. $2mrs - 2mr + ms - 4sr$
8. $\dfrac{1}{m} + \dfrac{1}{s} + \dfrac{1}{r}$
9. $\dfrac{3}{m} + \dfrac{2s}{r}$

4.6   Directed numbers

It is possible to explain what directed numbers are by referring to Fig. 4.1.

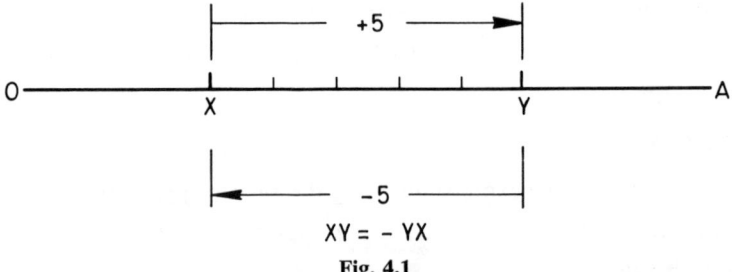

$$XY = -YX$$

Fig. 4.1

OA is any line upon which two points X and Y are marked, X and Y being 5 units apart. If the distance from X to Y is measured, i.e., from left to right, it is said to be a positive distance $+5$. If the distance is measured from Y to X, i.e., from right to left, it is a negative distance -5. We have two directions to consider, and in order to distinguish between them we use $+$ and $-$ signs with the numbers. Such numbers are called **directed numbers.** Thus

$$XY = +5$$
$$YX = -5$$

so that

$$XY = -YX$$

84

Directed numbers can be considered in another way. If a man possesses 10 pence he is said to have $+10$ pence. If he owes 10 pence he is said to have -10 pence. Directed numbers are thus seen to represent opposites. A third illustration of directed numbers may be taken from our ideas regarding time. Time forward is taken as positive and time back as negative. From a particular day 5 days later is $+5$, whilst 7 days earlier is -7.

Note: If a number does not have a sign, such as $7x$, it is understood to be $+$, that is $+7x$. This statement implies that all numbers in algebra are directed numbers.

(a) Addition and subtraction of directed numbers

The addition and subtraction of directed numbers may be illustrated by the diagrams in Fig. 4.2 relating to the five examples:

(a) $6+4$
(b) $6-4$
(c) $6-9$
(d) $-4+9$
(e) $-3-4$

In each case the first number in the calculation is represented by the line OX in the diagram, the second number by XY, and the answer by the line OY.

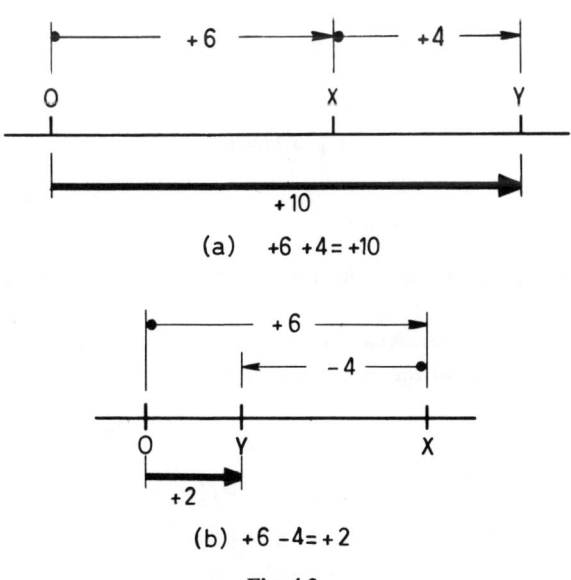

Fig. 4.2

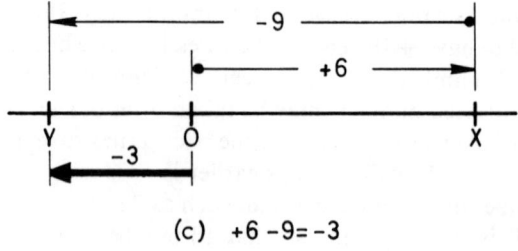

(c)   +6 −9= −3

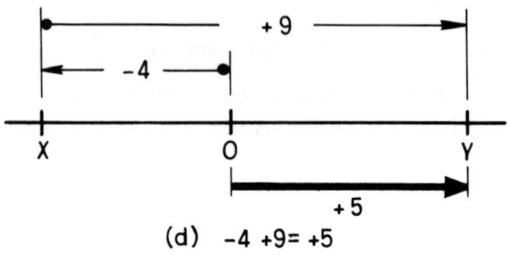

(d)   −4 +9= +5

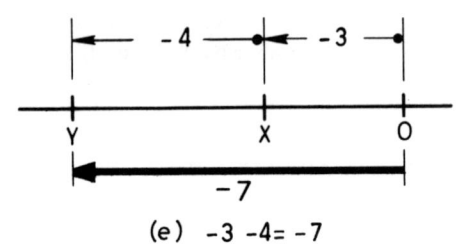

(e)   −3 −4= −7

Fig. 4.2 (*cont.*)

In order to avoid drawing diagrams every time the above results can be obtained by the following rules.

Rule 1. *If two numbers have the same sign they are added together, and the answer takes the common sign.*

Rule 2. *If two numbers have different signs they are subtracted, and the answer takes the sign of the larger number.*

For example:

$$-7-8 = -15$$

$$+2+9 = +11$$

$$-6+8 = +2$$

$$7-10 = -3$$

The two rules may be used to extend the problem to more than two numbers, as explained in Example 4.5.

EXAMPLE 4.5 Simplify: $-8t+7t+5t-6t-4t$.

Collecting the negative and positive terms together, the problem reduces to

$$-8t-6t-4t+7t+5t = -18t+12t$$
$$= -6t$$

(b) Addition and subtraction of negative numbers

Addition and subtraction of negative numbers of the type

$$6+(-5)$$

and

$$6-(-5)$$

can be illustrated using Fig. 4.1.

Since $$(-5) = YX$$

then

$$-(-5) = -YX = XY = +5$$

and

$$+(-5) = +YX = YX = -5$$

Therefore

$$6+(-5) = 6-5 = 1$$

and

$$6-(-5) = 6+5 = 11$$

As a working rule we can state that, for any two numbers M and N

$$M+(-N) = M-N$$
$$M-(-N) = M+N$$

EXAMPLE 4.6 In the following:

(a) $(-6)-(-7) = -6+7 = 1$
(b) $\quad 7-(-2) = \quad 7+2 = 9$
(c) $\quad 4-(+10) = 4-10 = -6$
(d) $\quad -7+(-4) = -7-4 = -11$
(e) $\quad 6+(-9) = -6-9 = -3$
(f) $\quad 5x-(-6x) = 5x+6x = 11x$
(g) $-3s-(-7s) = -3s+7s = 4s$
(h) $\quad 8q+(-4q) = 8q-4q = 4q$

(c) Multiplication and division of directed numbers

The rules of multiplication and division may be obtained by considering the velocity of a point P along a line XY in Fig. 4.3.

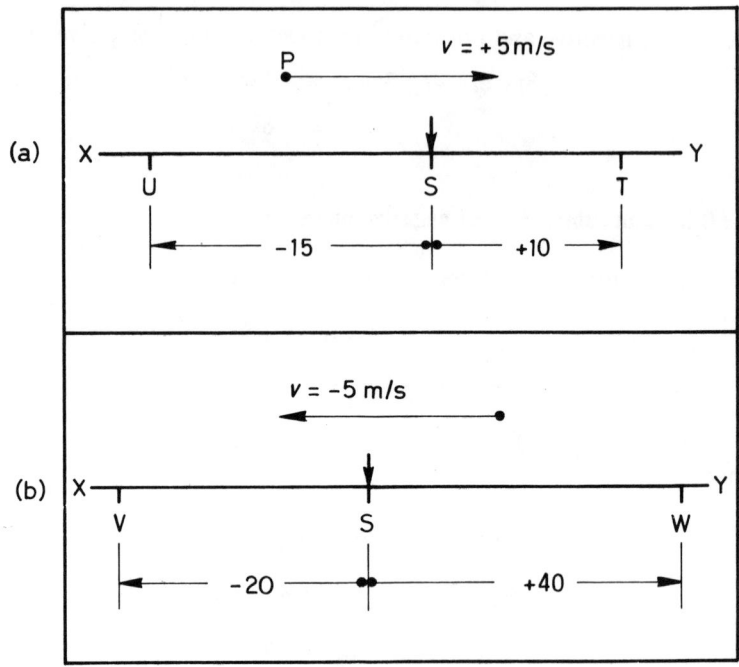

Fig. 4.3

Let P have a velocity $v = 5$ m/s, and at a particular instant let it be at the position S.

Let P move from left to right, as in Fig. 4.3(a), i.e., $v = +5$ m/s.
2 s later (i.e., $+2$), P will be at T, $+10$ m from S,

$$(+5) \times (+2) = +10 \qquad (1)$$

3 s earlier (i.e., -3), P was at U, -15 m from S,

$$(+5) \times (-3) = -15 \qquad (2)$$

Let P move from right to left, as in Fig. 4.3(b), i.e., $v = -5$ m/s.
4 s later, (i.e., $+4$), P will be at V, -20 m from S,

$$(-5) \times (+4) = -20 \qquad (3)$$

8 s earlier (i.e., -8), P was at W, $+40$ m from S,

$$(-5) \times (-8) = +40 \qquad (4)$$

From these four cases we can write down a rule for the multiplication of directed numbers.

88

Rule 1. *Multiplication of like signs gives* +, *multiplication of unlike signs gives* −.

From these results one can deduce that the division of numbers obeys the same rule.

Rule 2. *Division of like numbers gives* +, *division of unlike numbers gives* −.

EXAMPLE 4.7   In the following:

(a) $(-7) \times (-6) = +42$
(b) $(+9) \times (-3) = -27$
(c) $(14) \div (-7) = -2$
(d) $(-8) \div (-2) = +4$

EXERCISE 4.7

Collect the numbers in the following expressions:
1. $-9 + 6$
2. $-4 - 3$
3. $-7x + 9x - 14x + 10x$
4. $-14m + 2m + 3m$
5. $-9t - 7t - 14t$
6. $(-6) \times (-4)$
7. $(-4) \times (+5)$
8. $(+8) \times (-7)$
9. $(+10) \times (-2)$
10. $(-3y)(-4)$
11. $(-8)(-5s)$
12. $(-21) \div (7)$
13. $(21) \div (-7)$
14. $(-28) \div (-4)$
15. $(-8m) \div (4)$
16. $(21r) \div (-3)$

Simplify and collect the following expressions:

17. $-4 - (-6)$
18. $-(-5) + (-7)$
19. $+(-4) - (+8)$
20. $-(2z) - (-4z) + (7z)$
21. $-(-5p) - (15p) + (-2p)$
22. $5a - (-6a) + (-8a) - (+2a)$

4.7 Indices

In Section 2.4 we found that certain multiplications are written in the form

$$10 \times 10 = 10^2$$

or

$$10 \times 10 \times 10 = 10^3$$

Numbers other than 10 were also written in the same way, e.g.,

$$4 \times 4 = 4^2$$

$$6 \times 6 \times 6 \times 6 = 6^4$$

The same system is also applied to algebraic numbers, so we have:

$$x \times x \times x \times x \times x \times x = x^4$$

$$x \times x \times x \times x \quad = x^3$$

$$x \times x \quad\quad = x^2$$

$$x \quad\quad\quad = x^1, \quad \text{written as } x$$

It is useful at this point to compare indices with coefficients. We have seen already that:

$$x + x + x + x = 4x$$

$$x + x + x \quad = 3x$$

$$x + x \quad\quad = 2x$$

$$x \quad\quad\quad = 1x, \quad \text{written as } x$$

(a) Multiplication of numbers in index form

If the two numbers x^2 and x^3 are to be multiplied they must first be written out in full, i.e.,

$$x^2 \times x^3 = x \times x \times x \times x \times x = x^5$$

This result may be obtained by adding the two indices:

$$x^2 \times x^3 = x^{2+3} = x^5$$

This method may be applied to the multiplication of all numbers with the same base, and thus avoids the need for writing each number out in full.

The rule for multiplication may be stated in the following way:

$$x^n \times x^m = x^{n+m}$$

90

Rule. *The indices are added when the numbers are multiplied.*

In most cases these numbers with indices will have coefficients as well. Their multiplication is shown in Example 4.8.

EXAMPLE 4.8   Evaluate:

(a) $-3r^5 \times 2r^4$

(b) $4y^2 \times -3y^3 \times 2y^5$

   (a) Using the associative law

$$-3r^5 \times 2r^4 = -3 \times 2 \times r^5 \times r^4$$

$$= -6r^9$$

   (b) In this example the multiplication is extended to three terms. The coefficients and signs are multiplied first and then the y's by adding the indices, i.e.,

$$4y^2 \times -3y^3 \times 2y^5 = -4 \times 3 \times 2y^{2+3+5}$$

$$= -24y^{10}$$

(b) Division of numbers in index form

As we already know, division in arithmetic can often be carried out by cancelling common factors between numerators and denominators. Exactly the same thing can be done with algebraic numbers. For instance,

$$\frac{a \times a \times a \times a \times a}{a \times a \times a} = a \times a = a^2$$

If we write out this division in the form with indices it becomes,

$$\frac{a^5}{a^3} = a^2$$

The result is obtained by subtracting the indices, that is,

$$x^n \div x^m = x^{n-m}$$

Rule. *The indices are subtracted when the numbers are divided.*

In general these numbers will have coefficients as well. Their division is carried out as in Example 4.9.

EXAMPLE 4.9   Evaluate $-15d^9 \div -10d^4$.

$$-15d^9 \div -10d^4 = +\tfrac{15}{10}d^{9-4}$$

$$= \tfrac{3}{2}d^5$$

(c) The meaning of a number raised to the power 0

Following the rule obtained above we have

$$x^3 \div x^3 = x^0$$

It is necessary to find exactly what is meant by x^0. Writing the division out in full, we have

$$\frac{x \times x \times x}{x \times x \times x}$$

which is obviously 1. Therefore, it may be concluded that

$$x^0 = 1$$

Rule. *Any number raised to the power 0 is equal to* 1.

(d) Power of a power

By a power of a power we mean a number such as

$$(y^7)^2$$

In order to find what this number is, it is written out in full:

$$y^7 \times y^7 = y^{14}$$

It is seen that this result may be obtained by multiplying the indices, i.e.,

$$(y^7)^2 = y^{7 \times 2} = y^{14}$$

Therefore

$$(x^m)^n = x^{mn}$$

Rule. *The indices are multiplied together for the power of a power.*

EXAMPLE 4.10   In the following:

(a) $(f^4)^7 = f^{28}$
(b) $(2m)^3 = 2m \times 2m \times 2m = 2^3 m^3 = 8m^3$
(c) $(4g^5)^3 = 4^3 g^{15} = 64g^{15}$

EXERCISE 4.8

1. Write down the following in index form:

    (a) $x \times x \times x \times x \times x$
    (b) $3 \times 3 \times 3 \times 3$
    (c) $2y \times 2y \times 2y \times 2y$
    (d) $x^2 \times x^2 \times x^2 \times x^2 \times x^2$

2. Write down the following in full:

    (a) a^2 (b) $(x^2)^3$ (c) $(2x)^2$

3. Multiply out the following, giving the result in index form:

    (a) $p^4 \times p^5$
    (b) $2m^4 \times 3m^5$
    (c) $-3t^2 \times 6t^4$
    (d) $-5y^2 \times -4y^3$
    (e) $2x^4 \times 3x^5 \times 5x^2$
    (f) $-3x \times -4x^2 \times 5x^3$
    (g) $4f^3 \times -3f^2 \times 4f^4$

4. Divide the following:

    (a) $a^7 \div a^7$
    (b) $-27f^4 \div 3f^2$
    (c) $-16e^6 \div -12e^2$
    (d) $3q^5 \div -5q^4$

5. Simplify the following:

    (a) $(x^2)^3$ (b) $(-2v^4)^5$ (c) $(7w^3)^2$ (d) $(5^0)^4$ (e) $(3t^0)^4$

(e) Negative indices

A negative index may be understood from the following problems.

$$x^3 \div x^b = x^{3-b} = x^{-2}$$

but

$$x^3 \div x^5 = \frac{x \times x \times x}{x \times x \times x \times x \times x} = \frac{1}{x^2}$$

Thus

$$x^{-2} = \frac{1}{x^2}$$

Hence we see that a negative index is the reciprocal. Therefore, we have

$$x^{-3} = \frac{1}{x^3}$$

$$x^{-1} = \frac{1}{x}$$

$$5^{-2} = \frac{1}{5^2}$$

Note: When the number goes from the numerator to the denominator the sign of the index changes. Similarly it can apply when the index is positive. Thus,

$$x^3 = \frac{1}{x^{-3}}$$

$$6 = \frac{1}{6^{-1}}$$

(f) Fractional indices

From Section 4.7(d) we have

$$(x^6)^{\frac{1}{2}} = x^3$$

but

$$\sqrt{(x^6)} = x^3$$

so that the index $\frac{1}{2}$ represents the square root. Thus by an extension of the reasoning,

$$z^{\frac{1}{3}} = \sqrt[3]{(z)}$$

EXAMPLE 4.11   Simplify:

(a) $(9x^2)^{\frac{1}{2}}$

(b) $9(x^2)^{\frac{1}{2}}$

(c) $64^{\frac{1}{3}}$

(d) $25^{-\frac{1}{2}}$

(e) $\sqrt[3]{(64z^9)}$

(f) $125^{-\frac{2}{3}}$

(g) $(27t^6)^{\frac{2}{3}}$

(a) The bracket shows that the $\frac{1}{2}$ applies to the 9 as well as the x^2. Therefore

$$(9x^2)^{\frac{1}{2}} = 3x$$

(b) Here the $\frac{1}{2}$ does not apply to the 9, since the 9 is outside the bracket. Therefore

$$9(x^2)^{\frac{1}{2}} = 9x$$

(c) $(64)^{\frac{1}{3}} = (4 \times 4 \times 4)^{\frac{1}{3}} = (4^3)^{\frac{1}{3}} = 4$

(d) $25^{-\frac{1}{2}} = \dfrac{1}{25^{\frac{1}{2}}} = \dfrac{1}{5}$

(e) $\sqrt[3]{(64z^9)} = 4z^3$

(f) It will simplify the arithmetic if the cube root is calculated first:

$$(125)^{-\frac{2}{3}} = \frac{1}{(125)^{\frac{2}{3}}} = \frac{1}{(125^{\frac{1}{3}})^2} = \frac{1}{5^2} = \frac{1}{25}$$

(g) Here again the cube root is obtained first.

$$(27t^6)^{\frac{2}{3}} = (3t^2)^2 = 9t^4$$

EXERCISE 4.9

Evaluate the following:

1. $2x^{3/2} \times 2x^{5/2}$
2. $2x^{-1/4} \times 5x^{-5/4}$
3. $7t^{3/7} \times 2t^{-10/7}$
4. $t^{-6} \times t^{-4}$

5. $m^{-5} \div m^{-4}$
6. $p^{-3} \div p^5$
7. $12p^4 \div 8p^{-3}$
8. $3p^{3/4} \div 8p^{-1/4}$

Simplify the following:

9. $2(x^2)^2$
10. $(2x^2)^2$
11. $(27x^{12})^{1/3}$
12. $27(x^{12})^{1/3}$

13. $(64)^{-2/3}$
14. $\left(\dfrac{9}{16}\right)^{3/2}$
15. $(64)^{-5/6}$

4.8 Brackets

From the distributive law expressions such as $2(x+4)$ can be expanded as follows:

$$2(x+4) = 2 \times x + 2 \times 4$$
$$= 2x + 8$$

More complicated expressions involving directed numbers can be expanded in the same way, as shown in Example 4.12.

EXAMPLE 4.12 Remove the brackets in the following expressions:

(a) $-2(2x-7)$
(b) $3(12x-4) - 5(3x-6)$

(a) $-2(2x-7) = -2 \times 2x - 2(-7)$.

$$= -4x + 14$$

(b) In this problem each of the brackets are worked out separately:

$$3(12x-4) - 5(3x-6) = 3 \times 12x + 3(-4) - 5 \times 3x - 5(-6)$$
$$= 36x - 12 - 15x + 30$$
$$= 36x - 15x - 12 + 30$$
$$= 21x + 18$$

Now and again the brackets containing an expression will have no number outside, such as

$$(9t+4s) \quad \text{and} \quad -(5a-7b)$$

95

In each case we assume that the number outside is 1. Thus in the first case, on removing the brackets, we have:

$$(9t+4s) = 1 \times 9t + 1(+4s)$$
$$= 9t + 4s$$

In the second case, because of the $-$ sign, we must multiply by -1:

$$-(5a-7b) = -1 \times 5a - 1(-7b)$$
$$= -5a + 7b$$

It is possible to obtain expressions with far more complicated arrangements of brackets. In order to remove such systems of brackets we use the following rule.

Rule. *Remove the innermost brackets first, then the others in turn from the innermost to the outermost. At each point the expression is simplified.*

EXAMPLE 4.13   Remove the brackets and simplify the expression:

$$3[2(4x-3)-2(-x+7)]$$

Step 1. Leave the [ ] brackets intact and remove the inner ( ) brackets, giving

$$3[2 \times 4x + 2(-3) - 2(-x) - 2(7)] = 3[8x - 6 + 2x - 14]$$

Step 2. Collect and simplify the terms inside the brackets. The expression becomes $3[10x-20]$.

Step 3. Remove the outer brackets, to give

$$3 \times 10x + 3(-20) = 30x - 60$$

EXERCISE 4.10

Remove the brackets and simplify the expressions:

1. $3(a+2b+3c)$
2. $-3(x-4)$
3. $-(2a-b-c)$
4. $2(3s-4t)-5(s-2t)$
5. $-(5a+6b-c)+2(-3a+4b+c)$
6. $-3[5x-6(4x+7)-2(-x+4)]$
7. $(x^3+2x^4)-(x^4+x^3)-x^2$
8. $2[2(x-x^2+3x^3)-4(x^2-3x^3)]$
9. $-(-3a+b-2c)$

4.9 Multiplication of binomials

In the previous section the contents of a bracket were multiplied by a single number or letter, in accordance with the distributive law. Now the procedure is extended to the multiplication of the contents of a bracket by another bracket containing a binomial expression, such as,

$$(4x-3) \times (x-7)$$

The second bracket is multiplied by each term in the first, that is,

$$(4x-3)(x-7) = 4x(x-7) - 3(x-7)$$
$$= 4x^2 - 28x - 3x + 21$$
$$= 4x^2 - 31x + 21$$

EXERCISE 4.11

Multiply out or expand the following:

1. $(x+2)(x+1)$
2. $(y-3)(y+4)$
3. $(2z-6)(z-2)$
4. $(-x+5)(3x+7)$
5. $(x+y)(x-2y)$
6. $(a-b)(a+5b)$
7. $(8+y)(8+y)$
8. $(x+7)^2$
9. $(y+2u)^2$
10. $(x-y)(x+y)$
11. $(v-4u)(v+4u)$
12. $(6t-7s)(9t+4s)$

4.10 Factors

The process of factorization is the name given to the process of putting a number into products. In arithmetic $6 = 3 \times 2$, the 3 and 2 are called factors. In algebra, since numbers are more complicated, there are several ways of obtaining factors. The simplest are of the type

$$xy = x \times y$$

where x and y are the two factors. Again

$$3ab = 3 \times a \times b$$

where 3, a, and b are factors.

Type 1. One common factor to all terms

(a) Consider the expression

$$px+py$$

p is common to both terms. Take the common factor p outside the bracket and divide each term by the common factor p. Thus, it becomes

$$p(x+y)$$

The two factors are p and $(x+y)$. It is important to take out the highest common factor as seen in Example 4.14.

EXAMPLE 4.14 Factorize (p^2y+p^2z).

The highest common factor to both terms is p^2. Thus it becomes

$$p^2(y+z)$$

The three factors are p, p, and $(y+z)$.

EXAMPLE 4.15 Factorize $3ax+12ay+15az$.

The highest common factor is $3a$, giving

$$3a(x+4y+5z)$$

EXAMPLE 4.16 Factorize $4x^4+6x^3+10x^2$.

The highest common factor is $2x^2$. Thus the expression becomes

$$2x^2(2x^2+3x+5)$$

(b) It is often convenient to take -1 as a factor. For instance, if we have

$$-x+y$$

we can take out -1 as a factor to give

$$-1(x-y)$$

In reality, there is no need to write in the number 1, i.e., the expression can be written as

$$-(x-y)$$

This process is illustrated in Example 4.17.

EXAMPLE 4.17 Write out the following expressions with -1 as a factor:

(a) $-2p+3q$
(b) $-x^2-4x+7$

   (a) $-2p+3q = -(2p-3q)$

   (b) $-x^2-4x+7 = -(x^2+4x-7)$

98

(c) It often happens that the common factor is a binomial or a trinomial. For instance, the common factor in

$$3x(x+2)-4(x+2)$$

is the binomial $(x+2)$. Thus the expression factorizes to

$$(x+2)(3x-4)$$

It sometimes happens that a binomial is a common factor, apart from signs, as shown in Example 4.18.

EXAMPLE 4.18   Factorize $a(x-y)+b(-x+y)$.

We see that the two binomials have completely opposite signs in each letter, i.e., x and $-x$, $-y$ and y. The first step is to take -1 as a factor in the second term, so that the expression becomes

$$a(x-y)-b(x-y) = (x-y)(a-b)$$

Type 2. Grouping of terms in pairs with common factors

Consider the expression

$$mp - mq + np - nq$$

The first pair have a common factor m, the second pair a common factor n. These pairs are grouped together to give

$$(mp-mq)+(np-nq) = m(p-q)+n(p-q)$$
$$= (p-q)(m+n)$$

EXAMPLE 4.19   Factorize:

(a) $3x-9y+sx-3sy$
(b) $x^2-2x+4x-8$
(c) $z^2-4z-5z+20$

The factors become

(a) $(3x-9y)+(sx-3sy)$ $= 3(x-3y)+s(x-3y)$
$= (x-3y)(3+s)$

(b) $(x^2-2x)+(4x-8)$ $= x(x-2)+4(x-2)$
$= (x-2)(x+4)$

(c) $(z^2-4z)+(-5z+20)$ $= z(z-4)+5(-z+4)$
$= z(z-4)+5 \times -1(z-4)$
$= z(z-4)-5(z-4)$
$= (z-4)(z-5)$

EXERCISE 4.12

1. Factorize the following by looking for a single common factor:
   - (a) $ax+3ay+4az$
   - (b) $3t^2-9t^3$
   - (c) $mnx+mny$
   - (d) $ft-gt$
   - (e) a^2x-a^2y
   - (f) $ra-rb-rc$
   - (g) $5-5p$
   - (h) $4t^4+32t^3+8t^2$

2. Take out -1 as a factor in the following:
   - (a) $(-x+y)$
   - (b) $(-3z+4y-7x)$
   - (c) $3x-4$

3. Factorize the following:
   - (a) $x(4-z)+y(4-z)$
   - (b) $x(x+1)+4(x+1)$
   - (c) $l(4m+n)-4(4m+n)$
   - (d) $4l(x-6)-3(-x+6)$
   - (e) $4m(m+3)+3(-m-3)$
   - (f) $4t(x-y)+(x-y)$

4. Factorize the following by grouping in pairs:
   - (a) $ab-ac+eb-ec$
   - (b) $ut-vt-us+vs$
   - (c) $5xa+25ya-5xb-25yb$
   - (d) $x^2-6x+3x-18$
   - (e) $t^2-7t-4t+28$
   - (f) $x^2-ax-x+a$

Assessment test 4

1. In the number $8t^5$ write down
   - (a) the coefficient
   - (b) the base
   - (c) the power

2. Which of the following answers is the correct expansion of $-2(3x-4)$?
   - (a) $-6x-8$
   - (b) $-6x+8$
   - (c) $+6x+8$
   - (d) $+6x-8$

3. Each of the following calculations gives an answer of 24. Complete each answer by putting in the correct sign.

(a) $(-6) \times (-4)$
(b) $(-8) \times (3)$
(c) $(-72) \div (-3)$
(d) $(-120) \div (5)$

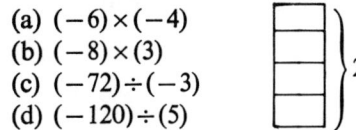 $\left. \right\}$ 24

4. List I contains four calculations. List II gives the answers. Match the correct answers to the calculations by filling in the appropriate numbers in the boxes.

List I	List II
A. $-6-4$	1. $\quad 2$
B. $-6+4$	2. -10
C. $+6-4$	3. 10
D. $+6+4$	4. $\quad -2$

A	B	C	D

5. State which is the correct answer to $7x + 8x - 5y$.

(a) $10xy$
(b) $7 + 3xy$
(c) $15x - 5y$
(d) $2xy + 8x$

6. Each of the following calculations gives an answer of $14t^2$. Complete the answer by filling in the correct sign in each case.

(a) $(-7t) \times (-2t)$
(b) $(+7t) \times (-2t)$
(c) $(28t^3) \div (-2t)$
(d) $(42t^3) \div (3t)$

$\left. \right\}$ $14t^2$

7. Evaluate $3 \times 2a + 7a \times 5 - 6a \div 2$.

8. Fill in the boxes with the correct numbers.

(a) $3x^3 \times 2x^2 = \Box x^{\Box}$
(b) $12x^6 \div 3x^{\Box} = \Box x^4$
(c) $(16m^{16})^{\frac{1}{4}} = \Box m^{\Box}$

(d) $2x^{-3} = \dfrac{2}{\Box x^{\Box}}$

(e) $(\Box z^2)^{\Box} = 8z^6$
(f) $(2z^2)^0 = \quad \Box$
(g) $2(z^2)^0 = \quad \Box$

9. Which of the following are the correct pair of factors of $-px+py-pz$? More than one pair may be correct.
   (a) $p,\ (-x+y-z)$
   (b) $p,\ (x+y-z)$
   (c) $-p,\ (x-y+z)$
   (d) $-p,\ (x-y-z)$

10. If $U = x/y$, choose **two** correct answers for U when $x = 20$ and $y = \frac{1}{2}$.
    (a) $\frac{2}{20}$
    (b) 10
    (c) 20×2
    (d) $20 \div \frac{1}{2}$

11. Which of the following gives an answer of 1?
    (a) $4x - x - 3x$
    (b) $2t^2 \div t \div 2t$
    (c) $(3y)^0$
    (d) y^{-1}
    (e) $\dfrac{(64)^{\frac{1}{2}}}{4 \times 2}$

12. Determine
    (a) $\sqrt[3]{(64x^6)}$
    (b) $\sqrt{(64x^6)}$
    (c) $\left(\frac{18}{32}\right)^{\frac{1}{2}}$
    (d) $\left(1\frac{11}{25}\right)^{\frac{1}{2}}$

13. Factorize $x(a-b)+(-a+b)$ and state which of the following is the correct expression:
    (a) $(a-b)(x+1)$
    (b) $(a-b)(x-1)$
    (c) $x(a-b)$
    (d) $(-a+b)(x+1)$

14. List I shows numbers to the power -1. List II gives the same numbers written with positive powers. Match List 2 to List 1 by filling in the appropriate numbers in the boxes.

List I | List II

A. $2x^{-1}$ 1. $2x$

B. $(2x)^{-1}$ 2. $\dfrac{2}{x}$

C. $\dfrac{1}{(2x)^{-1}}$ 3. $\dfrac{1}{2x}$

D. $\dfrac{1}{2x^{-1}}$ 4. $\frac{1}{2}x$

A	B	C	D

15. Complete the following statements:
    (a) Indexed numbers are multiplied by .... the indices, if they have the same base.
    (b) Indexed numbers with the same base are .... by subtracting the indices.
    (c) A number raised to the power .... is equal to 1.
    (d) The square root of a number will have an index of ....

16. If $\dfrac{1}{x}$ is greater than $2y$, write $>$

    $\dfrac{1}{x}$ is smaller than $2y$, write $<$

    $\dfrac{1}{x}$ is equal to $2y$, $=$

    when x and y have the following values:
    (a) $x = 5,\ y = \frac{1}{10}$
    (b) $x = \frac{4}{5},\ y = \frac{4}{3}$
    (c) $x = \frac{1}{2},\ y = \frac{1}{2}$
    (d) $x = 3,\ y = \frac{1}{6}$

17. The laws of algebra are labelled as follows:
    Commutative law   X
    Distributive law   Y
    Associative law   Z

State which law each of the following obeys, by filling in X, Y or Z in the box.

(a) $yzx \times 3 = 3xyz$
(b) $5x + 2y - 3x = 5x - 3x + 2y$
(c) $-(3x - 2y) = -3x + 2y$
(d) $3x \times 2y = 6xy$

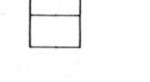

18. By first factorizing the expression, simplify $\frac{z}{2} + \frac{z}{3} + \frac{z}{6}$.

19. By removing the brackets state which of the following is the correct simplification of $2[(2x + y) + (-2x - y)]$:

(a) $8x - 4y$
(b) $8x$
(c) 0
(d) $4y$

5. Algebra 2

Objectives

After working through this chapter you should be able to

1. Recognize equations and expressions.
2. Solve simple equations.
3. Solve simple equations involving brackets and fractions.
4. Construct and solve simple equations relating to practical problems in science and engineering.
5. Solve simultaneous equations.
6. Evaluate formulae from given data.
7. Transpose formulae which are used in engineering and science.

5.1 Expressions and equations

In Section 4.1 the word 'term' was defined and explained. A series of terms is called an **expression**.

A single term such as $3x$ is called a **monomial** expression. Two terms, such as $3x+2y$ is called a **binomial** expression. Three terms, such as $3x+2y-4z$ is called a **trinomial** expression.

Any number can be substituted for these letters and the value of the expression determined. For example, when

$$x = 1, \quad y = 2, \quad z = 3, \quad 3x+2y-4z = 3 \times 1 + 2 \times 2 - 4 \times 3 = -5$$

or

$$x = 0, \quad y = 1, \quad z = 1, \quad 3x+2y-4z = 3 \times 0 + 2 \times 1 - 4 \times 1 = -2$$

An **equation** can be explained in terms of balance.

In engineering, we are familiar with a weighing balance. When a correct balance is obtained the weights in either scale pan are the same. We say that the weights in either scale pan are equal.

This concept of balance also occurs in algebra where it is often found that two expressions are equal. For example, we may find that the two expressions:

$$3x+3y \quad \text{and} \quad 4x-7y$$

are equal. To show that the two are equal they are written down with an equal sign ($=$) between them, i.e.,

$$3x+3y = 4x-7y$$

This is called an **equation**.

Consider now the equation

$$y = 4x + 7$$

We cannot substitute any pair of values for x and y. For instance, if $x = 3$, $y = 2$, the left-hand side $= 2$ and the right hand side $= 4 \times 3 + 7 = 19$. With these values, the equation does not balance. Therefore, the substitution $x = 3$, $y = 2$ is not admissible.

The value that y can have *depends* on the value that is allocated to x. For example, if $x = 1$, $y = 4 \times 1 + 7 = 11$. Hence, $x = 1$, $y = 11$ is an admissible substitution, since both sides now balance.

The letter x, because it can be given any value, is called an **independent variable**.

The letter y, since it depends on x, is called the **dependent variable**.

The admissible substitution is called the **solution** of the equation.

5.2   Simple equations

If the equation contains a single variable, such as x, having a power of 1, the equation is called a **simple** equation. The equation

$$2x = 10$$

is a simple equation. It can be seen that the *only* value of x which will make the equation balance is $x = 5$, that is

$$2 \times 5 = 10$$

This conclusion will apply to all simple equations.

A simple equation will have only one solution.

The method of solving a simple equation can be explained using the balance shown in Fig. 5.1.

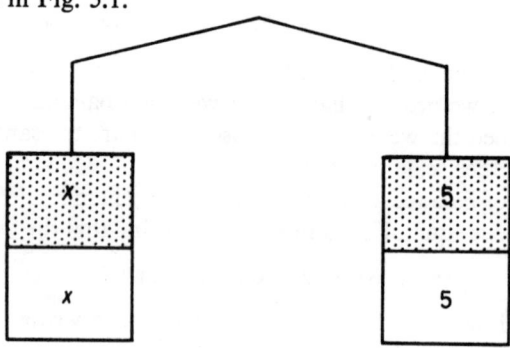

Fig. 5.1

On the left is seen the quantity $2x$, and on the right the quantity 10. If the left is divided by 2 we obtain x. If the right-hand side, likewise, is halved, we obtain 5. These quantities are shown by the shaded regions, which are seen to be equal, that is

$$x = 5$$

Therefore, it is seen that an equation can be modified in any way, provided that any change made on one side is the same as that made on the other side.

Using this principle simple equations can be solved without using diagrams, as shown in Example 5.1. The aim is to get the variable on its own on the left-hand side.

EXAMPLE 5.1   Solve the equation $7x - 4 = 31$.

Step 1. Add 4 to both sides.

$$7x - 4 + 4 = 31 + 4$$
$$7x \qquad = 35$$

Step 2. Divide both sides by 7:

$$\frac{7x}{7} = \frac{35}{7}$$
$$x = 5$$

Having obtained the solution a check should be made to see that it is correct. The solution is substituted into the left-hand side, and if necessary to the right-hand side, when the resulting two values should be the same. Thus

$$\text{Left-hand side} = 7 \times 5 + 4 = 31$$
$$\text{Right-hand side} \qquad = 31$$

A more complicated equation is rearranged, such that the terms containing the variable are collected on the left and the numbers on the right. This is illustrated in Example 5.2.

EXAMPLE 5.2   Solve the equation:

$$7y - 14 = 2y - 6 - 3y + 16$$

Step 1. Simplify both sides as much as possible:

$$7y - 14 = 2y - 3y - 6 + 16$$
$$= -y + 10$$

Step 2. Add 14 to both sides, and simplify,

$$7y - 14 + 14 = -y + 10 + 14$$

that is,

$$7y = -y + 24$$

Step 3. Add y to both sides, and simplify,

$$7y + y = -y + y + 24$$

that is,

$$8y = 24$$

Step 4. Divide both sides by the coefficient of y,

$$\frac{8y}{8} = \frac{24}{8}$$

and therefore

$$y = 3$$

This solution may be checked by substituting it back into the original equation.

Check. Substitute $y = 3$ into the equation.

Left-hand side is: $\qquad\qquad 7 \times 3 - 14 = 21 - 14 \qquad = 7.$

Right-hand side is: $\quad 2 \times 3 - 6 - 3 \times 3 + 16 = 6 - 6 - 9 + 16 = 7$

Therefore, $y = 3$ produces balance so that it must be the correct solution.

EXERCISE 5.1

Solve the following equations:
1. $3x = 12$
2. $4y = 9$
3. $3x - 7 = 17$
4. $2x - 3 = -4x + 27$
5. $20 - 3s + 3 = 2s - 12$
6. $9m + 3 - 7m = -4m - 17$
7. $-5t + 8 - 4t = 5t - 13 + 14$
8. $6x + 21 - 6x + 14 = -6x - 16$

5.3   **More difficult simple equations**

More difficult equations may be obtained containing brackets and fractions.

Simple equations containing brackets

If we were required to solve an equation such as

$$3(x-4)+4x = 2(2x+5)-1$$

the brackets would have to be removed first.

Step 1. Remove the brackets and simplify both sides:

$$3x-12+4x = 4x+10-1$$
$$7x-12 = 4x+9$$

Step 2. Add 12 to both sides and simplify:

$$7x-12+12 = 4x+9+12$$
$$7x = 4x+21$$

Step 3. Subtract $4x$ from both sides:

$$7x-4x = 4x+21-4x$$

that is,

$$3x = 21$$

Therefore

$$x = 7$$

Check in the original equation.

Left-hand side is: $\quad 3(7-4)+4(7) = 3(3)+4(7) = 9+28 = 37$

Right-hand side is: $\quad 2(2\times7+5)-1 = 38-1 = 37$

$x = 7$ produces balance, so that $x = 7$ is the correct solution.

Simple equations involving fractions

If the solution of

$$\frac{3x-4}{3}-\frac{3x}{4} = \frac{1}{6}$$

were required, the first step would be to remove the fractions. The LCM of the denominator is 12.

Step 1. Multiply throughout by the LCM:

$$\frac{12(3x-4)}{3}-\frac{12\times3x}{4} = \frac{12\times1}{6}$$

Step 2. Cancel out the denominators:

$$4(3x-4)-3\times3x = 2$$

109

Step 3. Remove the brackets and simplify:

$$12x - 16 - 9x = 2$$
$$3x - 16 = 2$$
$$3x - 16 + 16 = 2 + 16$$
$$3x = 18$$
$$x = 6$$

EXERCISE 5.2

Solve the equations:

1. $\frac{2}{3}(5x-1) - \frac{4}{5}(2x-3) = 0$

2. $\frac{3}{5}(4z-7) + 8 = 2z$

3. $7(4t-5) - 5t = 5(3t+1)$

4. $\dfrac{3y+7}{2} - \dfrac{5y-4}{3} = 0$

5. $\frac{5}{2}t - 12 = 7 + \frac{4}{3}t$

6. $\dfrac{w+2}{2} - \dfrac{w+8}{4} = \dfrac{w+3}{5}$

7. $\frac{1}{4}(2x-1) + \frac{1}{5}(4x+5) = \frac{1}{20}x + 2$

8. $\dfrac{3x}{4} = \dfrac{x}{2} - \dfrac{6-x}{3}$

5.4 Construction and solution of equations in engineering and science

There are many practical problems in engineering and science, which can be solved by constructing a simple equation. The equation so formed is then solved following the method shown in Section 5.3. One or two typical examples are given below.

EXAMPLE 5.3   A beam supports three loads. The first load is four times the second, and the third is 100 kN. If the sum of the three loads is 700 kN, find the values of the first two.

Let x be the second load. Then the first load is $4x$.

$$\text{The total load} = 4x + x + 100 = 700$$

Therefore

$$5x = 600$$

$$x = \frac{600}{5}$$

$$\text{The second load} = 120 \text{ kN}$$

$$\text{The first load} = 4x = 4 \times 120$$

$$= 480 \text{ kN}$$

EXAMPLE 5.4   A wire, 32 cm long, is to be bent into a shape of a rectangle, whose length and breadth are in the ratio 5 : 3. Find the length and breadth.

Let the length be $5x$ and the breadth be $3x$.

$$\text{Perimeter} = 5x + 3x + 5x + 3x = 32$$

$$16x = 32$$

$$x = 2$$

Therefore

$$\text{length} = 5 \times 2 = 10 \text{ cm}$$

and

$$\text{breadth} = 3 \times 2 = 6 \text{ cm}$$

EXERCISE 5.3

1. A firm produces two milling machines. One costs twice as much as the other. Both together cost £1590. Find the cost of each machine.
2. A rectangle has a width which is 3 m less than the length. Its perimeter is 22 m. Calculate the length and width.
3. Three consecutive numbers add up to 72. Find the three numbers.
4. A man bought a second-hand car for £x. He paid a further £$\frac{1}{4}x$ on repairs. Road tax and insurance were £75. His total outlay was £390. How much did he pay for the car?
5. Two men X and Y, produce components for an engine. X finds that working at a certain rate he has to reject 4 components per hour. Y works at $\frac{3}{4}$ the rate of X and rejects, on the average 1 component per hour. At the end of an 8 h day X has 10 more serviceable components than Y. Find the rate of working of X.

6. A current of 16 A divides into three branch networks. Four times the current flows in the first branch as compared with the second; 6 A flow in the third branch. Find the current in each of the first two branches.

7. An electric light bulb left on continuously was found to last 10% more than the guaranteed life. A similar bulb frequently switched on and off was found to last 10% less. There was a difference of 200 h between the lives of the two bulbs. Calculate the guaranteed life.

8. The original price of a certain tool doubled, and then increased by a further £7, to reach a final value of £25. What was the original price of the tool?

9. In one rotation of a grinding wheel, $M \times 10^{-6}$ kg of metal is removed. Find the number of rotations required to remove 2×10^{-3} kg.

10. Ball bearings are taken from the store to replace worn ones in three machines. Machine A requires 4 times as many replacements as machine B, and machine C requires 20 replacements. If 95 ball-bearings were taken from the store how many were required in machines A and B?

11. A spring extends 2 cm under its own weight. A load M causes a further extension. A load N causes an extension 3 times as much as M. The total extension due to M, N, and the weight of the spring together is 14 cm. What is the extension due to load N?

5.5 Simultaneous equations

When an equation contains only one unknown, such as one of the equations in Section 5.2, it is called a simple equation. An equation such as

$$x + 3y = 7$$

contains two unknowns, x and y. A whole range of values of x and y exist which satisfy this equation. If we have another equation

$$x + y = 3$$

this equation, too, will have a whole range of values. However, there will be only one value of x and one value of y, which will satisfy both equations. The two values are called the solutions to the pair of equations. The pair of equations together are called **simultaneous equations**.

The steps for solving the equations are as follows:

Step 1. Make the coefficients of one unknown, i.e., x or y, the same in both equations.

Step 2. Eliminate this unknown by *adding or subtracting* the equations to produce a simple equation in the other unknown.

Step 3. Substitute back into one of the original equations to obtain a simple equation in the first unknown.

Step 4. Make a check by substituting the two values into the equation *not* used in step 3.

EXAMPLE 5.15 Solve the equations:

$$2x + 5y = 10$$

$$x + 2y = 3$$

Step 1. Make coefficients of x the same by multiplying the second equation by 2:

$$2x + 5y = 10 \tag{1}$$

$$2x + 4y = 6 \tag{2}$$

Step 2. Subtract: $\qquad\qquad\qquad y = 4$

Step 3. Substitute in (1): $\qquad 2x + 20 = 10$

$$2x = -10$$

$$x = -5$$

The solution is $\qquad\qquad x = -5, \quad y = 4$

Step 4. Check in the second equation $x + 2y = 3$:

$$\text{Left-hand side} = -5 + 2(4) = 3$$

$$\text{Right-hand side} \qquad\quad = 3$$

EXAMPLE 5.6 Four lathes of type A and ten lathes of type B turn out 450 components per week. Two lathes of type A and eight lathes of type B turn out 270 components per week. Find out the number of components turned out by each type of lathe per week.

Let lathe A turn out x components per week, and lathe B turn out y components per week. Therefore

$$4x + 10y = 450 \tag{1}$$

$$2x + 8y = 270 \tag{2}$$

Multiply equation (2) by 2:

$$4x + 10y = 450 \tag{1}$$

$$4x + 16y = 540 \tag{3}$$

Subtract (1) from (3)

$$6y = 90$$

$$y = 15$$

113

Substitute into (1)

$$4x + 150 = 450$$

$$4x = 300$$

$$x = 75$$

Check in equation (2)

$$2(75) + 8(15) = 270$$

EXERCISE 5.4

Solve the following simultaneous equations:

1. $x + y = 4$
    $x - y = 2$

2. $x + 2y = 8$
    $x - 2y = 4$

3. $3x + y = 2$
    $2x + 3y = 5$

4. $3x + 2y = 10$
    $4x - y = 6$

5. $2p - 5q = 0$
    $-8p + 3q = 17$

6. $x + 2y = 3$
    $2x - 3y = 13$

7. $x + 2y = 6$
    $4x + 3y = -1$

8. $3E + 4I = -4$
    $-2E - 3I = 2$

5.6 Evaluation of formulae

Engineering and science use many formulae to express certain quantities in terms of other variables and constants. The values of these quantities can then be found when actual values are assigned to the variables. The following is a list of just a few of these formulae,

$v = u + at$ for the final velocity of a body

$\dfrac{1}{R} = \dfrac{1}{R_1} + \dfrac{1}{R_2}$ for the equivalent resistance R of a circuit

$T = 2\pi \sqrt{\dfrac{L}{g}}$ for the period of a simple pendulum

The values of v, R, T can be determined by substituting values for the variables in the respective formulae.

EXAMPLE 5.7   The final velocity of a car is given by

$$v = u + at$$

Find v, when $u = 30$ m/s, $a = 3$ m/s^2, $t = 5$ s.

Substituting these values into the equation gives

$$v = 30 + 3 \times 5$$
$$= 30 + 15$$
$$= 45 \text{ m/s}$$

EXAMPLE 5.8   The equivalent resistance R of a parallel circuit is

$$\frac{1}{R} = \frac{1}{R_1} + \frac{1}{R_2}$$

Find R, when $R_1 = 30$ ohm, $R_2 = 40$ ohm.

Substituting these values into the equation gives

$$\frac{1}{R} = \frac{1}{30} + \frac{1}{40}$$

$$- \frac{4+3}{120}$$

$$= \frac{7}{120}$$

Therefore, inverting both sides of the equation,

$$R = \frac{120}{7} = 17\tfrac{1}{7} \text{ ohm}$$

EXERCISE 5.5

1. The period of a simple pendulum is given by $T = 2\pi\sqrt{(L/g)}$. Find T when $L = 4.9$ and $g = 10$.
2. The kinetic energy E of a body is given by $E = \tfrac{1}{2}mv^2$. Find E when $m = 5$ kg and $v = 12$ m/s.
3. The final length L of a rod, after heating through a temperature t, is given by $L = l(1 + \alpha t)$. Find L when $l = 50$, $t = 80$, and $\alpha = 0.00002$. Find also the difference $(L - l)$.

4. The potential drop across the terminals of a battery is given by

$$V = \frac{Er}{R+r}.$$

Find V when $E = 2$, $r = 0.5$, and $R = 7$. Find also the difference $(E-V)$.

5. The velocity of a particle moving a distance s is given by $v^2 = u^2 + 2as$. Find v when $u = 9$, $a = 5$, and $2s = 4$.

6. The current in a circuit is given by

$$I = \frac{E}{R+r}.$$

Find the value of I when $E = 12$, $R = 13$, and $r = 2$.

7. The formula relating object and image distances v, u to the radius of curvature R of a curved mirror is

$$\frac{1}{u} + \frac{1}{v} = \frac{2}{R}$$

Find R when $v = 20$ and $u = -10$.

8. The heat supplied to a container of mass m containing water of mass M is given by $Q = (ms + MS)t$. Find Q when $m = 100$, $M = 1500$, $s = 0.4$, $S = 4.0$, and $t = 6$.

9. The resistance r of a wire increases to R when it is heated through a temperature t, the formula being $R = r(1 + \alpha t)$. Find R when $r = 60$, $\alpha = 0.0003$, and $t = 60$.

10. The heat developed in a wire is given by $H = I^2 Rt$. Find H when $I = 2$, $R = 12$, and $t = 5$. State what happens to H, when I is doubled, R is halved, and t is halved.

5.7 Transposition of formulae

In Section 5.6, examples of formulae used in engineering and science were given, in which one variable, called the dependent variable, is expressed in terms of other variables. Such a formula is $v = u + at$, where v is the dependent variable. In transposition, the formulae is modified in order to make one of the other letters the dependent variable. *In order to do this, any change on one side of the equation must be matched with an identical change on the other side of the equation.* The process of finding a new dependent variable is sometimes called **changing the subject**.

(a) Formulae not containing roots or powers

Consider a formula of the type

$$L = MN + P$$

in which N is to be made the new dependent variable. Three steps can be listed to carry out the process.

Step 1. Change the equation around if necessary so that the side containing the new subject (N in this case) is on the left, i.e.,

$$MN+P = L$$

Step 2. Rearrange the equation so that the *term* containing the new subject is on its own. In our example we wish to have the term MN on its own, so P is subtracted from both sides:

$$MN+P-P = L-P$$

that is,

$$MN = L-P$$

Step 3. Finally, obtain the subject on its own, i.e., divide both sides by M:

$$\frac{MN}{M} = \frac{L-P}{M}$$

that is,

$$N = \frac{L-P}{M}$$

EXAMPLE 5.9 The total surface area of a cylinder is given by

$$A = 2\pi r(r+h)$$

Express h in terms of A and r.

Step 1. $\qquad\qquad 2\pi r(r+h) = A$

Step 2. Divide both sides by $2\pi r$:

$$r+h = \frac{A}{2\pi r}$$

Step 3. Subtract r from both sides:

$$h = \frac{A}{2\pi r} - r$$

(b) Equations containing roots

Consider the equation $\sqrt{P} = 3$. This can be written as $P^{\frac{1}{2}} = 3$. In order to find P, both sides of the equation must be raised to the power 2, that is

$$(P^{\frac{1}{2}})^2 = 3^2$$

$$P = 9$$

When an equation containing a root has to be transposed and the new subject is inside the root, the root must be removed by raising both sides of the equation to the appropriate power.

EQUATION 5.9 If $A = \sqrt{(x-y)}$ find the expression for x.

Since x, the new subject is inside the root, both sides of the equation must be raised to the power 2 in order to transpose it.

$$A^2 = [(x-y)^{\frac{1}{2}}]^2 = x-y$$

$$x-y = A^2$$

$$x-y+y = A^2+y$$

$$x = A^2+y$$

(c) Equations containing powers

Consider the equation $M^3 = 125$. To find M the cube root of both sides of the equation must be obtained, that is,

$$(M^3)^{\frac{1}{3}} = (125)^{\frac{1}{3}}$$

$$M = 5$$

When an equation containing a power has to be transposed and the new subject is under this power, then an appropriate root must be taken.

EXAMPLE 5.10 If $v^2+u^2 = 2gs$ find an expression for v.

$$v^2+u^2 = 2gs$$

Obtain v^2 on its own on the left-hand side

$$v^2+u^2-u^2 = 2gs-u^2$$

$$v^2 = 2gs-u^2$$

Take square roots on both sides.

$$\sqrt{(v^2)} = \sqrt{(2gs-u^2)}$$

$$v = \sqrt{(2gs-u^2)}$$

(d) Equations containing fractions

First, multiply throughout by the LCM of the denominators.

EXAMPLE 5.11 If

$$\frac{(x+y)}{y} = \frac{4M}{N},$$

find the equation for N.

The LCM is yN. The equation is multiplied by yN:

$$\frac{yN(x+y)}{y} = \frac{yN4M}{N}$$

Therefore

$$N(x+y) = 4yM$$

Divide both sides by $(x+y)$

$$N = \frac{4yM}{(x+y)}$$

EXERCISE 5.6

1. The perimeter of a rectangular door is $P = 2a + 2b$. Make b the subject of this equation.
2. The area of a trapezium is $A = (a+b)h$. Transpose making a the subject.
3. Transpose the following, making z the subject in each case:
   (a) $a + 2z = b$
   (b) $2abz = c$
   (c) $3 - 2bz = x + y$
   (d) $az = 6 - bz$

   (e) $\dfrac{r}{z} = (x+y)$

4. The sum of a series is $S = a/(1-r)$. Express r in terms of the other quantities.
5. The velocity equation is $v = u + at$. Rearrange and obtain a in terms of the other quantities.
6. The period of a simple pendulum is $T = 2\pi\sqrt{(l/g)}$. Express l in terms of the other quantities.
7. Energy is given by $E = \frac{1}{2}mv^2$. Transpose, giving v as the subject.
8. Young's modulus is $Y = Fl/Ax$. Transpose, making l the subject.
9. The volume of a sphere is $V = \frac{4}{3}\pi r^3$. Find the formula for r.
10. The velocity in simple harmonic motion is $v = w\sqrt{(a^2 - y^2)}$. Rearrange to obtain the formula for y.
11. A formula for the drop in voltage in a battery is

$$V = \frac{Er}{R+r}$$

   Make r the subject.
12. A rod of length l expands to a length L when heated, where L is given by

$$L = l(1 + \alpha T)$$

   α being the coefficient of expansion. Find α in terms of the other quantities.

13. The moment of area of a ring is

$$M = \pi(R^4 - r^4)$$

Express r in terms of the other letters.

Assessment test 5

1. State whether each of the following is an expression or an equation:
   (a) $4x - y = 31$
   (b) $4x - y - 31$
   (c) $\dfrac{8}{x} = 7 + 3x$
   (d) $8x$

2. Starting with the equation $y = 3x + 2$, which of the following equations are obtained from it?
   (a) $2y = 6x + 4$
   (b) $\dfrac{y}{3} = x + 2$
   (c) $y - 3x = 2$
   (d) $y + 2 = 3x$

3. Which of the following equations has a solution $x = 1$?
   (a) $3x = 3$
   (b) $3x = x + 3$
   (c) $2(x - 1) = 3(x - 1)$
   (d) $x + 1 = -x + 1$

4. If $5x = 20$, which of the following gives the true value of x?
   (a) $x = \frac{5}{20}$
   (b) $x = \frac{20}{5}$
   (c) $x = 20 - 5$
   (d) $x = 20 \times 5$

5. If $x + 3 = 7$, which of the following gives the true value of x?
   (a) $x = \frac{7}{3}$
   (b) $x = 7 + 3$
   (c) $x = 7 - 3$
   (d) $x = 3 - 7$

6. If $V = \pi r^2 h$ state which of the following equations are obtained from it?
   (a) $\dfrac{V}{r} = \pi r h$

(b) $h - V = 2\pi r^2 h$

(c) $r^2 = \dfrac{V}{\pi h}$

(d) $V + \pi r^2 h = 2\pi r^2 h$

7. If $3x + 1 - 4(2x - 1) = 0$, which of the following is the solution for x?
   (a) 5
   (b) -5
   (c) 1
   (d) -1

8. If
$$x + y = 10$$
$$x - y = \;\; 3$$
find x and y.

9. If
   (a) $2x + 5y = 3$
   (b) $3x - 4y = 7$
   which of the following gives the correct method to solve the equations?
   (i) Multiply (a) by 3 and (b) by 2 and add
   (ii) Multiply (a) by 3 and (b) by 2 and subtract
   (iii) Multiply (a) by 4 and (b) by 5 and add
   (iv) Multiply (a) by 4 and (b) by 5 and subtract.

10. If $R = r + \dfrac{1}{r}$ List I gives the values of R for the values of r given in List II.

    Match the correct values of r and R by filling in the appropriate number in the empty boxes.

List I	List II
A. $5\frac{1}{5}$	1. 1
B. 2	2. $\frac{1}{2}$
C. $2\frac{1}{2}$	3. $\frac{1}{10}$
D. $10\frac{1}{10}$	4. 5

A	B	C	D

11. If
$$\frac{x}{y} = \frac{y+2}{x}$$
which of the following equations are obtained from it?

(a) $\dfrac{y}{x} = \dfrac{x}{y+2}$

(b) $\dfrac{x^2}{y} = (y+2)$

(c) $\dfrac{x^2}{y} = (y+2)x$

(d) $\dfrac{x}{y} - (y+2) = \dfrac{1}{x}$

12. If $v = u + at$, which of the following is the correct equation for t?

    (a) $t = \dfrac{v-a}{u}$

    (b) $t = \dfrac{u-v}{a}$

    (c) $t = \dfrac{v-u}{a}$

    (d) $t = \dfrac{a}{v-u}$

13. If $x = 2$, fill in the missing gaps denoted by a $*$, in the following:

    (a) $x + * = 11$

    (b) $5x = *$

    (c) $\dfrac{x}{4} + * = 3\frac{1}{2}$

    (d) $* \times x - 7 = 3$

14. List I gives four formulae for V. List II gives the same equations with h as the subject. Match the correct pairs of equations by filling in the correct numbers in the vacant boxes.

	List I		List II
A.	$V = \pi r^2 h$	1.	$h = V - \pi r^2$
B.	$V = \frac{1}{3}\pi rh$	2.	$h = \dfrac{V}{\pi r^2}$
C.	$V = \pi r^2 + h$	3.	$h = \dfrac{V - r^2}{\pi}$
D.	$V = \pi h + r^2$	4.	$h = \dfrac{3V}{\pi r}$

A	B	C	D

15. If $s^2 = 25$, which of the following is correct?
    (a) $s = 25^2$
    (b) $s = \frac{1}{2} \times 25$
    (c) $s = 25^{\frac{1}{2}}$
    (d) $s = 25^{-2}$

6. Diagrams and graphs

Objectives

After working through this chapter you should be able to

1. Convert data from one system of units into a related system.
2. Plot the sets of related data on a pair of parallel axes.
3. Relate corresponding pairs of values by a third line, called a link.
4. Explain what is meant by one-to-one mapping.
5. Define and name two axes at right angles.
6. Determine suitable scales for the axes.
7. Plot accurately a point with given co-ordinates.
8. Draw a straight line to fit points which satisfy a linear law.
9. Read values from a straight-line graph.
10. Determine the gradient of a straight line.

6.1 Conversion of data into a related system of units

In engineering and science there are many formulae relating different quantities. For example,

$$\text{Mass } M = \text{Density } D \times \text{Volume } V$$

or

$$\text{Voltage } E = \text{Current } I \times \text{Resistance } R.$$

Such relationships are called equations.

Consider the first equation applied to a material whose density is 2 g/ml, that is $D = 2$, so that

$$M = 2V$$

Written in this way, M is the dependent variable and V is the independent variable. For any value of V the mass of the material can be calculated using the above equation. To each value of V there is a particular value of M. These pairs of values are called *corresponding* values, and they are related by the equation. Thus a table of corresponding values can be obtained by simple substitution, as shown below.

Volume V (ml)	1	2	3	4	5
Mass M (g)	2	4	6	8	10

Sets of related data such as this can be represented pictorially using two
axes which are:
  (a) Parallel,   or
  (b) Perpendicular.

6.2   Representation of related values using two parallel axes

The above table of related values can be represented on two parallel axes,
as shown in Fig. 6.1.

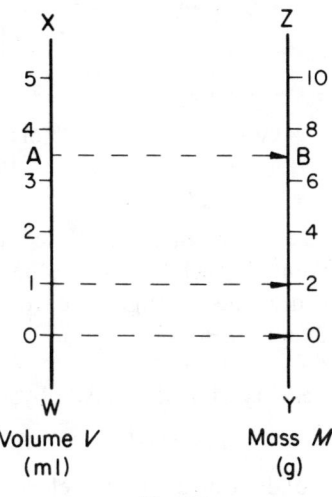

Fig. 6.1

On the axis WX is marked, to scale, some values of V, and likewise, on the
axis YZ, some values of M. The scale on YZ is twice the scale on WX.
Corresponding values are shown by the dotted lines. These may be called
link lines. These link lines are all parallel in this example, so that the mass of
any volume A can be found by drawing a parallel line AB.

Making the link lines parallel restricts the choice of scales on the two axes,
so that it is necessary to consider link lines which are not parallel. Consider
the case where the material has a density of 7 g/ml, when

$$M = 7V$$

The table of related values is

Volume V (ml)	1	2	3	4	5
Mass M (g)	7	14	21	28	35

These values are shown on the axes in Fig. 6.2.

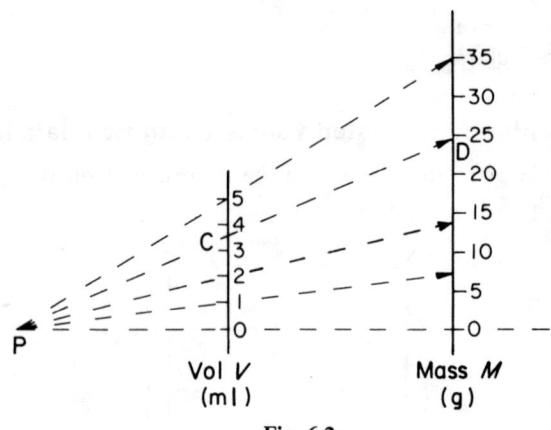

Fig. 6.2

Corresponding pairs are again given by the dotted link lines. It is seen that they all intersect at the point P. Therefore, any other pair of corresponding values can be obtained by drawing a straight line from P to cut the axes at C and D. Thus when $V = 3\frac{1}{2}$ ml, $M = 24\frac{1}{2}$ g.

EXAMPLE 6.1   The velocity of a car is given by

$$v = 3t + 5$$

where t is the time in seconds. Calculate the velocity corresponding to the times 1 s, 2 s, 3 s, 4 s and 5 s. Plot these related values on two parallel scales, and draw link lines to relate corresponding values.

From the axes, find v when $t = 3\frac{1}{2}$ s and t when $v = 12\frac{1}{2}$ m/s.

The table of corresponding values is found by substituting the various values of t into the equation.

Time t (s)	1	2	3	4	5
Velocity v (m/s)	8	11	14	17	20

The related values are shown in Fig. 6.3. All the link lines meet at P. P can now be used to link corresponding pairs of values.

To find v when $t = 3\frac{1}{2}$, join P to A. The intersection at B gives the corresponding value $v = 15\frac{1}{2}$ m/s.

To find t when $v = 12\frac{1}{2}$, join P to D, and continue it to C where it is seen that $t = 2\frac{1}{2}$ s.

126

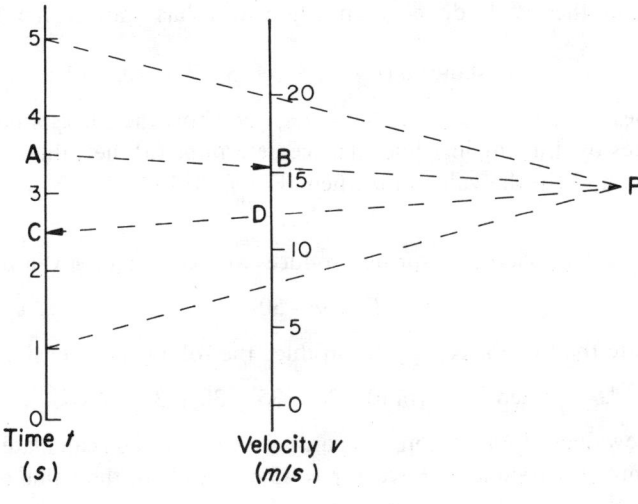

Fig. 6.3

EXERCISE 6.1

1. The distance travelled by a lorry is given by

$$d = 10t$$

where t is the time in seconds. Calculate the distances travelled in the following times:

Time t (s)   1,   2,   3,   4,   5

Plot the related values of d and t on two parallel axes. Draw link lines to corresponding values. Hence find (a) the value of d when $t = 4\frac{1}{2}$ s and (b) the value of t when $d = 35$.

2. The voltage E across a resistor of 3 ohm is given by

$$E = 3I$$

where I is the current in amperes. Calculate the values of E when the current is given as follows:

Current I (A)   1,   2,   3,   4,   5

Plot the related values of E and I on two parallel axes. Draw link lines to corresponding values. Hence find (a) the value of E when $I = 4\frac{1}{2}$ A and (b) the value of I when $E = 18$ V.

3. The work done W in Joules by a small wheel in moving a distance d metres against a force is

$$W = 120d$$

127

Calculate the work done when the wheel has moved the following distances,

Distance d (m) 2, 3, 5, 9, 10, 12

Plot these values on two parallel scales, and show the corresponding pairs of values by drawing link lines. Hence determine (a) the value of W when $d = 7$ m and (b) the value of d when $W = 1000$ J.

4. The force F applied to a spring produces an extension e given by

$$F = 5e - 50$$

Calculate the force necessary to produce the following extensions,

Extension e (mm) 20, 25, 30, 35, 40

Plot the values of F and e on two parallel axes, and then calculate from the diagram (a) the value of F when $e = 45$ mm and (b) the value of e when $F = 25$ N.

5. The heat supplied to a quantity of brass, for various temperature rises, is given by the following table:

Temperature T (°C)	2	3	4	5	6
Heat Q (J)	80	120	160	200	240

Plot these two sets of values on parallel axes. From the graph, if

$$Q = kT$$

find the value of k.

6. The force F acting on a body causes acceleration a. The corresponding values are given by the following table:

Acceleration a (m s^{-2})	4	6	8	10	12
Force F (N)	32	48	64	80	96

Plot these values on two parallel axes. If

$$F = ka$$

use your axes to determine k.

6.3 Mappings

In the previous section all the examples used are of the type that any value of one variable produces only one corresponding value of the other variable. In Example 6.1 for instance, for each value of t there is only one value of v. The relationship between v and t is called a **mapping**. The mappings are represented by the link lines. Since there is only one value of v to any value of t, the mapping is said to be **one to one**.

6.4 Two axes at right angles

The pairs of corresponding values obtained in Example 6.1 can be represented differently, by using two axes at right angles to one another. The vertical axis is labelled the y-axis. In this example it becomes the v-axis. The horizontal axis is called the x-axis. In this example, it becomes the t-axis. The two axes are shown in Fig. 6.4.

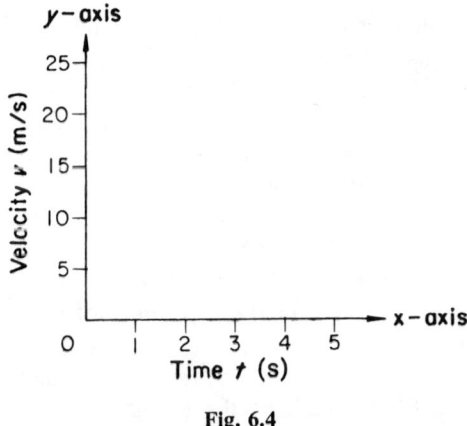

Fig. 6.4

The point of intersection of the two axes is called the **origin**. At the origin the value of both scales is zero. Scales are marked on the axes, as with the parallel axes, with the values increasing outwards from the origin. Such axes are called **cartesian axes**.

It is often necessary to deal with negative numbers when plotting values. We have already come across negative numbers in Chapter 4. Negative values of the x-axis are plotted to the left of the origin, and negative values of y below the origin, as shown in Fig. 6.5.

129

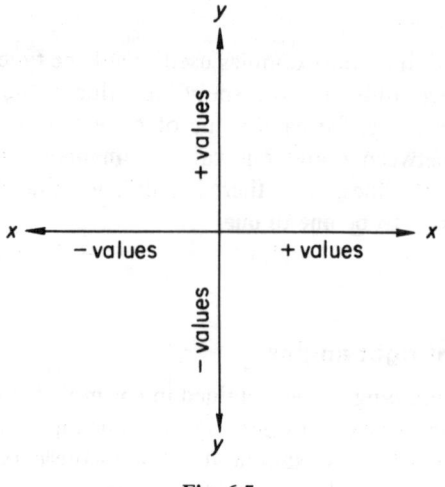

Fig. 6.5

6.5 Scales on the axes

The axes are normally drawn on graph paper, which is divided into large squares and small squares. The side of each large square is subdivided into 10 equal parts. In selecting a suitable scale for either axis, one consideration must be to make the plotting of data easy. In order to do this the scale on the large square must be such that the small squares are simple sub-multiples of it. Therefore, the large squares are usually given a scale of 1, 2, 5, 10, 20 units, etc., so that the corresponding value of each small square is $0·1$, $0·2$, $0·5$, 1, 2, etc.

The other important point to remember is that the data to be plotted should use most of the possible length of the axis, and not be confined to a small section of it. Badly selected scales are shown in Fig. 6.6. In Fig. 6.6(a) both the scales on both x- and y-axes are too large. In Fig. 6.6(b) the scale on the y-axis is satisfactory, but the scale on the x-axis is too big. In both instances the data to be plotted is bunched into the shaded area, with a resulting reduction in accuracy.

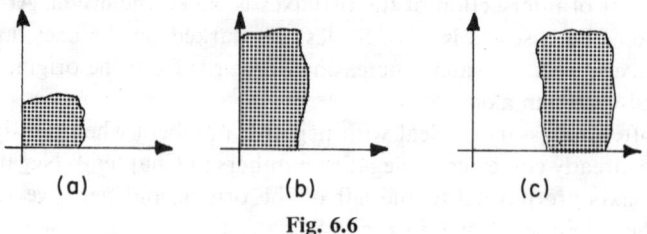

(a)   (b)   (c)

Fig. 6.6

Example 6.2 shows how to select a suitable scale for an axis.

EXAMPLE 6.2   The values of current used in an electrical circuit increases from 1 A to 30 A. There are 11 large squares on the axis. Select a suitable scale, and mark it on a diagram.

Step 1. Divide the maximum value of I by the number of squares. That is $\frac{30}{11} = 3$ approximately

Step 2. 3 is not a convenient scale, since it is difficult to subdivide it to fit the smaller squares.

Step 3. Choose the nearest suitable scale stated above that is greater than 3. The nearest scale is 5. Therefore, the most appropriate scale is 5 per large square, as shown in Fig. 6.7.

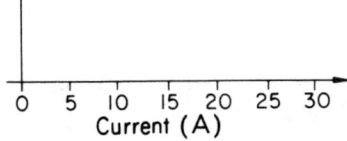

Fig. 6.7

Fig. 6.6(c) shows another example where bunching of data occurs. In this case the data do not include values near zero. Starting the scale at zero produces a situation where all the data are confined to a small section of the axis. It is then necessary to select a scale which does not start at zero, as shown in Example 6.3.

EXAMPLE 6.3   The current in a circuit varies from 18 A to 35 A. Select a scale suitable scale, if the axis has 11 large squares.

Following Example 6.2, a suitable scale appears to be $\frac{35}{11} = 3$ approximately. For ease of working this becomes 5 per large square. The scale is shown in Fig. 6.8(a), where the range from 18 to 35 is shaded. It is seen that bunching occurs in the region 18–35.

In this instance, since the data is not near to zero, it is better to *find* the range of values to be plotted, and to divide this by 11.

$$\text{Range} = 35 - 18 = 17 \text{ A}$$

$$\text{Scale} = \frac{17}{11} = 2 \text{ approx.}$$

Since this value makes subdividing easy, it can be adopted. Now the scale starts at 16 A, as shown in Fig. 6.8(b), and the data plotted use the whole scale.

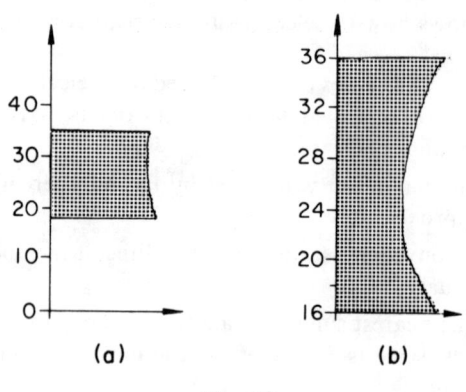

Fig. 6.8

EXERCISE 6.2

Select the most suitable scale for the following data, and draw the axis, marked with the scale in each case.

1. Time changes from 1 s to 12 s. Axis has 11 large squares.
2. Force changes from 3 N to 30 N. Axis has 8 large squares.
3. Mass changes from 4 kg to 16 kg. Axis has 7 large squares.
4. Heat input changes from 30 J to 35 J. Axis has 5 large squares.
5. Voltage changes from 27 V to 45 V. Axis has 8 large squares.
6. Diameter changes from 0·5 mm to 3·5 mm. Axis has 8 large squares.
7. Temperature changes from 45 °C to 55 °C. Axis has 9 large squares.
8. Temperature changes from 0 °C to −50 °C. Axis has 10 large squares.
9. Length increases from 100·0 mm to 101·7 mm. Axis has 9 large squares.
10. Resistance increases from 3·5 ohm to 3·9 ohm. Axis has 9 large squares.

6.6 Plotting a point given its co-ordinates

The corresponding values on the two parallel scales in Section 6.2 can now be represented on the two scales at right angles, shown in Fig. 6.4. From Example 6.1 consider the corresponding values $t = 1$, $v = 8$. Using the axes as drawn in Fig. 6.9, the value of $t = 1$ is shown at the point A. A dotted line is drawn vertically from A. Similarly, the point $v = 8$ is shown at the point B. A dotted line is drawn horizontally from B. These two dotted lines meet at the point R. R is a point in which $t = 1$ and $v = 8$. Therefore, the corresponding values are represented by a single point. These two corresponding values are called the **co-ordinates** of the point R. The point R is written with the co-ordinates (1, 8) after it. Note that the x co-ordinate is written first. Q is another point shown on the diagram, having the co-ordinates (4, 20).

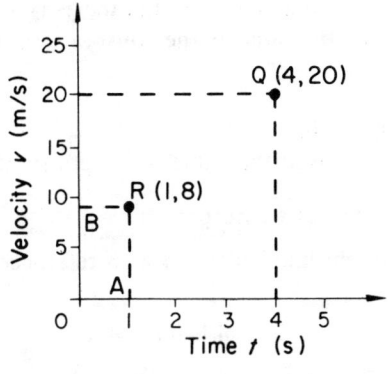

Fig. 6.9

EXERCISE 6.3

Plot the following points on Cartesian axes:

1. (2, 4), (3, 7), (5, 8), (6, 11), (5, 1), (0, 8), (8, 0)
2. (15, 20), (18, 35), (10, 10)
3. (−5, 7), (−6, −2), (4, −6), (−2, −8), (−3, 5)

6.7 Straight-line graphs

The corresponding pairs of values in Example 6.1 are now plotted on two axes at right angles, using the method described in Section 6.6. The result is shown in Fig. 6.10. It is seen that all the points representing the corresponding values lie on a straight line, ST. When data of this type lie on a straight line, the quantities involved obey a **linear** law and the equation is said to be a linear equation.

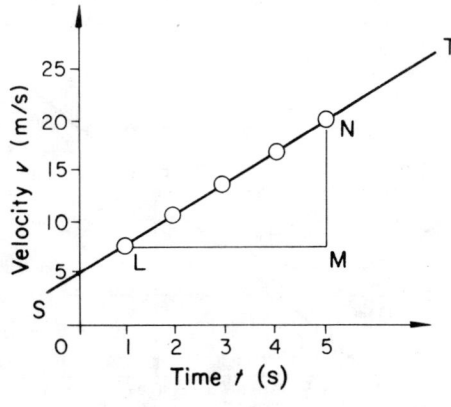

Fig. 6.10

It is interesting to note that the line cuts the y-axis at the point S where $v = 5$. This is seen to be the same as the constant 5 in the original equation

$$v = 3t + 5$$

This is always found to be the same.

The graph cuts the y-axis at the value of the constant in the equation.

6.8 Gradient of a straight-line graph

The gradient of a straight line is defined with reference to the graph in Fig. 6.10. It is the ratio

$$\frac{\text{vertical height MN}}{\text{horizontal distance LM}}$$

In calculating the gradient, the horizontal distance should be made a convenient value, if possible. In the example,

$$\text{Horizontal distance} = 5 - 1 \quad = 4$$

$$\text{Vertical height} = 20 - 8 = 12$$

$$\text{Gradient} = \tfrac{12}{4} \quad = 3$$

6.9 Values from a straight-line graph

In Example 6.1 corresponding values were obtained by drawing link lines such as AB. Corresponding values can also be obtained from straight-line graphs. Fig. 6.11 shows a straight-line graph of the temperature of a body plotted against time, from the following table.

Temperature T (°C)	10	14	18	22	26	30
Time t (s)	0	1	2	3	4	5

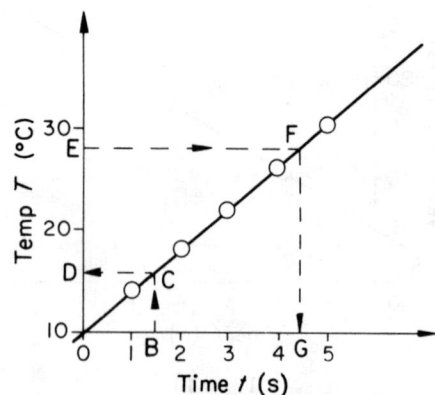

Fig. 6.11   Time t (s)

The Fig. 6.11 shows how to determine (a) the temperature of the body at a time $t = 1\frac{1}{2}$ s and (b) the time when the temperature is 28 °C.

The value $t = 1\frac{1}{2}$ s is shown as B on the x-axis. A dotted line BC is drawn up to the graph. At C a dotted line is drawn horizontally to the y-axis at D, where $T = 16$ °C, that is,

$$T = 16\,°C \quad \text{when} \quad t = 1\tfrac{1}{2}\,\text{s}$$

Similarly, the value $T = 28$ °C is shown at E. The dotted line EFG is drawn as shown, to meet the x-axis at G, when $t = 4\frac{1}{2}$ s, that is

$$t = 4\tfrac{1}{2}\,\text{s} \quad \text{when} \quad T = 28\,°C$$

EXERCISE 6.4

In each of the questions in Exercise 6.1 draw a straight-line graph to fit the data. From the graph calculate the values asked for in each question.

Determine, also, the gradient of each graph.

Assessment test 6

1. The table shows corresponding values of two related variables s and t. Complete the table, by filling in the values for s in the empty boxes.

t	1	2	3	4	5	6
s	7	10		16	19	

2. Corresponding values of related data are shown on the two axes in the diagram. Answer the questions below.

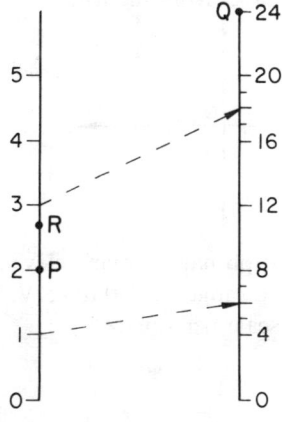

Fig. AT 6.1

(a) Draw the link line from P to its corresponding value.

(b) Mark the corresponding value of Q on the left-hand axis.

(c) Find the point of intersection of the two dotted link lines. Use this point to mark the point corresponding to R on the right-hand axis.

(d) Will the values found change if the two axes are moved further apart? Answer **yes** or **no**.

3. The two axes in the figure represent related data. From the figure determine the corresponding values of A, B, C, and D.

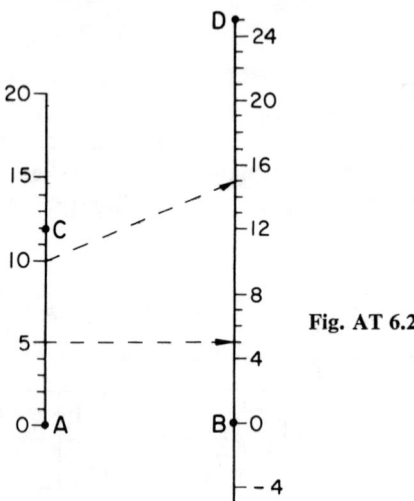

Fig. AT 6.2

4. The figure shows cartesian axes. Match the sections of axes by marking them with the correct label from the list.

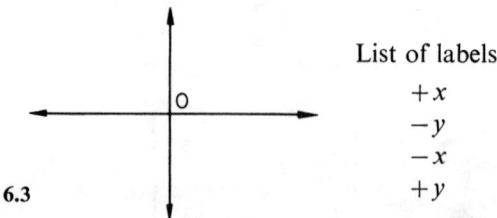

Fig. AT 6.3

List of labels

$+x$

$-y$

$-x$

$+y$

5. An axis is drawn on graph paper having eleven large squares. A voltage is to be plotted having a range from 0 to 45 V. Which of the following is the most appropriate scale per square?

(a) 4·5 V

(b) 5 V

(c) 4 V

(d) 6 V

6. An axis has five large divisions marked on it. It is used, in turn, to plot the data in List I. List II gives the appropriate scale per large division. Match the scales to the range of data by filling in the appropriate number in the boxes.

List I	List II
A. 0–20	1. 0·2
B. 25–35	2. 5
C. 7–8	3. 10
D. 5–55	4. 2

A	B	C	D

7. Referring to question 6, List III gives the starting values on the axis for each of the range of data in List I. Match the correct starting value to the range, by filling in the appropriate letter in the boxes.

List III
(a) 25
(b) 7
(c) 5
(d) 0

A	B	C	D

8. What are the co-ordinates of the point P shown in the diagram? Select one answer from the following:

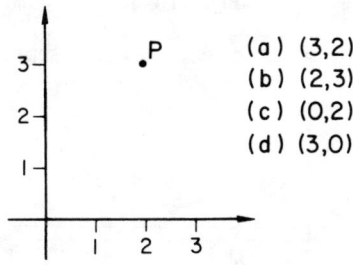

(a) (3,2)
(b) (2,3)
(c) (0,2)
(d) (3,0)

Fig. AT 6.4

9. A straight line is shown in the figure.

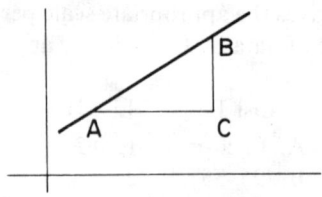

Fig. AT 6.5

What is the gradient? Select the correct answer from the following:

(a) $\dfrac{AC}{BC}$

(b) $AC \times BC$

(c) $\dfrac{BC}{AC}$

(d) AB

10. What is the gradient of the straight line shown in the figure?

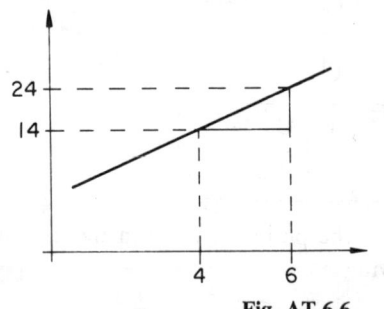

(a) $\dfrac{10}{2}$

(b) $\dfrac{2}{10}$

(c) $\dfrac{24}{6}$

(d) $\dfrac{4}{14}$

Fig. AT 6.6

11. Which of the following graphs **does not** represent a linear law?

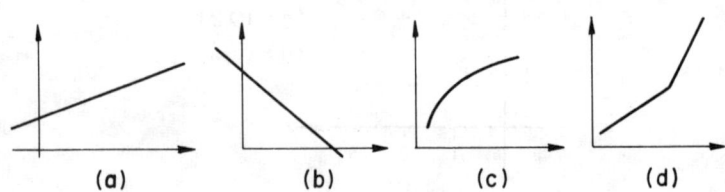

(a)   (b)   (c)   (d)

Fig. AT 6.7

138

12. Using the graph in the figure, read off the value of M which corresponds to the value $V = 3$, and the value of V which corresponds to the value $M = 20$.

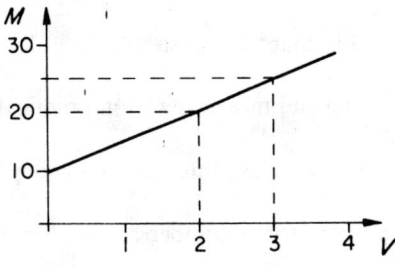

Fig. AT 6.8

7. Statistics

Objectives

After working through this chapter you should be able to

1. Collect data by counting and measuring from practical work in other level 1 subjects.
2. Explain the terms, sample, class interval, frequency, relative frequency, range.
3. Group data into an appropriate number of equal intervals.
4. Complete the tally diagram and a frequency table.
5. Draw the following labelled diagrams to show how class frequencies and relative frequencies vary within a set of data.
   (a) 100% bar chart
   (b) pie diagram
   (c) pictogram
   (d) horizontal and vertical bar chart
   (e) histogram.
6. Discuss the pattern of the data by examining the diagrams in 4 and 5.

7.1 Introduction

Statistics is the branch of mathematics that is concerned with the methods of collecting and handling numbers, called **data**, which have been obtained by measuring and counting.

Examples where such data have been collected are:
(a) measurement of the lengths of 100 steel bars,
(b) counting, by a factory inspector, of the number of transistors being packed in cartons by an automatic machine.

In the first example it is fairly easy to measure all the bars. In the second example, if the machine is working continuously it would not be very practical for the inspector to check more than about 10% of the cartons, that is, only a *sample* of the cartons may be chosen for checking purposes.

7.2 Display of data

After the data has been collected it must be displayed in an easily examined diagram to see what pattern the results follow. It then becomes possible to reach some conclusions and, if necessary, take some action. For instance, in example (b) if the intention is to pack 50 transistors in each carton, and

140

significantly more or less than 50 is actually being packed, it could mean that the machine needs resetting or overhauling.

It is not easy to detect a pattern in a large amount of numerical data, and to help in detecting any pattern it is necessary to display the information in a methodical way. To illustrate the method of display consider the following example, which gives to the nearest hour, the time taken by forty apprentices to complete a particular job.

```
65  57  37  46  78  56  32  51  54  58  27  68  56  45
52  51  47  52  64  50  59  54  40  58  61  57  39  65
53  63  50  72  49  83  57  55  62  46  74  53
```

It is seen that is is difficult to detect any pattern in this set of times. The data is then usually grouped into **classes** of equal **class interval**. A ciass interval in the above example is a certain number of hours, say 5. A class will then be, say, from 25 to 30 h or from 40 to 45 h. In practice, the size of the class interval is chosen so that the whole set is grouped into 6 to 15 classes. If the data is grouped into less than 6 or more than 15, it will still be difficult to detect any pattern. A suitable class interval can be found by first finding the **range** of the information. The range is the spread of data from the smallest value to the greatest value. In our example

$$\text{Range} = \text{greatest value} - \text{smallest value}$$

$$= 83 - 27$$

$$= 56 \text{ h}$$

If a class interval of 5 h is chosen there will be $\frac{56}{5}$ classes, that is 11 or 12 classes, which is a reasonable number. Suitable classes will therefore be 25–29 h inclusive, 30–34 h inclusive, etc., up to 80–84 h.

7.3 Frequency table

From the data a table is now drawn up showing the number of apprentices in each class. The number in each class is called the **class frequency** and the complete table is called a **frequency table**.

To help in the preparation of a frequency table a **tally diagram** is drawn up as the information is being counted. A tally mark represents one item from the data. As a further aid to counting every fifth tally is drawn across the previous four, as shown in the table below. The total number of tally marks in each class gives the frequency for that class.

141

Time in hours	Tally count	Frequency
25–29	1	1
30–34	1	1
35–39	11	2
40–44	1	1
45–49	⊥⊦⊦⊤	5
50–54	⊥⊦⊦⊤ ⊥⊦⊦⊤	10
55–59	⊥⊦⊦⊤ 1111	9
60–64	1111	4
65–69	111	3
70–74	11	2
75–79	1	1
80–84	1	1
	total	40

It will be seen from this table that the tally diagram and the frequency table both make it much easier to detect a pattern in the information. It appears that the majority of the apprentices are taking about 45–65 h to complete the task. Only 5 apprentices finish in less than 45 h, and only 7 take longer than 65 h.

EXERCISE 7.1

In each of the following examples use a tally diagram to complete the frequency table.

1. The wages in £ of 50 employees on a production line is as follows,

```
65  66  65  67  64  66  65  66  65  68
68  64  66  68  64  68  66  67  64  66
64  68  65  61  67  63  67  62  65  67
66  67  65  63  67  65  67  62  66  64
65  62  67  68  66  68  63  66  64  63
```

Use a class interval that gives an eight-class frequency table. Comment upon the wage pattern.

2. The marks obtained by 60 students in an examination were as follows:

```
60  74  30  64  80  65   7  20  22  42  15  53
46  70  30  13  63  80  38  10  27  38  55  50
21  33  37  32  47  73  40  21  88  25  35  34
29  39  54  56  48  90  69  15  12  45  19   5
27  43  75  62  49  72  52  24  39  26  36  57
```

Use ten classes in your table. Have you any comment to make about the examination?

3. An automatic machine is set to produce steel bars with a standard length of 31·5 cm. A sample of 40 bars gave the following results:

$$
\begin{array}{cccccccccc}
31\cdot4 & 31\cdot3 & 31\cdot4 & 31\cdot3 & 31\cdot4 & 31\cdot5 & 31\cdot3 & 31\cdot5 & 31\cdot4 & 31\cdot4 \\
31\cdot1 & 31\cdot4 & 31\cdot3 & 31\cdot5 & 31\cdot4 & 31\cdot4 & 31\cdot4 & 31\cdot4 & 31\cdot2 & 31\cdot3 \\
31\cdot3 & 31\cdot6 & 31\cdot5 & 31\cdot6 & 31\cdot5 & 31\cdot4 & 31\cdot4 & 31\cdot4 & 31\cdot4 & 31\cdot4 \\
31\cdot4 & 31\cdot5 & 31\cdot4 & 31\cdot4 & 31\cdot4 & 31\cdot3 & 31\cdot2 & 31\cdot4 & 31\cdot5 & 31\cdot5
\end{array}
$$

From your frequency table how do you think the production process is going?

4. A sample of 50 ball bearings produced by a machine had the following diameters:

$$
\begin{array}{cccccccccc}
16\cdot4 & 15\cdot5 & 15\cdot8 & 16\cdot0 & 15\cdot8 & 16\cdot2 & 16\cdot0 & 16\cdot2 & 15\cdot9 & 16\cdot1 \\
15\cdot8 & 16\cdot1 & 16\cdot0 & 16\cdot1 & 15\cdot9 & 15\cdot7 & 16\cdot1 & 15\cdot5 & 15\cdot8 & 15\cdot9 \\
16\cdot1 & 16\cdot4 & 16\cdot1 & 16\cdot2 & 16\cdot2 & 16\cdot1 & 16\cdot0 & 16\cdot0 & 15\cdot9 & 16\cdot0 \\
15\cdot8 & 16\cdot0 & 15\cdot5 & 16\cdot1 & 16\cdot0 & 15\cdot9 & 16\cdot0 & 16\cdot1 & 15\cdot9 & 16\cdot1 \\
16\cdot6 & 16\cdot6 & 16\cdot1 & 16\cdot1 & 16\cdot2 & 16\cdot3 & 16\cdot3 & 16\cdot3 & 16\cdot3 & 16\cdot2
\end{array}
$$

Arrange the information in six equal classes.

7.4 Pictorial displays

It has been seen that a tally diagram and a frequency table make it easier to examine large amounts of data. It will be found easier still to detect a pattern if the results are displayed in picture form.

The following pictorial methods may be used:
(a) 100% bar chart
(b) pie diagram
(c) pictogram
(d) horizontal and vertical bar charts
(e) histogram

Before discussing these methods it is necessary to define another term used in statistics, namely **relative frequency**.

The relative frequency is the proportion of the total frequency that is contained in any particular class. In our example the actual frequency in the 50–54 h class is 10. The total frequency for the sample is 40.

$$\text{Relative frequency for 50–54 h class} = \frac{\text{class frequency}}{\text{total frequency}}$$

$$= \tfrac{10}{40} = \tfrac{1}{4}$$

This can also be expressed as a percentage, that is,

$$\text{Relative frequency} = \tfrac{1}{4} \times 100$$

$$= 25\%$$

The total relative frequency for the sample is the sum of all the relative frequencies in each class, which must be 100%.

(a) 100% bar chart

A bar chart is used to represent the distribution of the relative frequencies for the sample. The length of the bar is subdivided into lengths proportional to the relative frequencies. The total length of the bar represents the total relative frequency of 100%. This method of representation is suitable for a set of results containing few classes.

EXAMPLE 7.1 A survey in an engineering firm showed that apprentices travelled to work by the following means, bus 12, car 15, on foot 4, train 8, pedal cycle 10, motor cycle 11. Represent the information on a 100% bar chart.

Mode of travel	Frequency	% Relative frequency
Bus	12	$\frac{12}{60} \times 100 = 20$
Car	15	$\frac{15}{60} \times 100 = 25$
On foot	4	$\frac{4}{60} \times 100 = 6\cdot7$
Train	8	$\frac{8}{60} \times 100 = 13\cdot3$
Pedal cycle	10	$\frac{10}{60} \times 100 = 16\cdot7$
Motor cycle	11	$\frac{11}{60} \times 100 = 18\cdot3$
total	60	100

The 100% bar chart is shown in Fig. 7.1.

Fig. 7.1 100% bar chart

(b) Pie chart

With a pie chart a circle is divided into sectors, the sizes of the sectors being proportional to the relative frequencies. The size of a sector is directly proportional to the angle contained between its radii. Therefore, to draw a pie chart, the 360° angle at the centre of the circle must be divided in proportion to the relative frequencies. Consider this being carried out for the data in Example 7.1.

Mode of travel	Frequency	% Relative frequency	Angle of sector
Bus	12	20	20% of 360 = 72°
Car	15	25	25% of 360 = 90°
On foot	4	6·7	24°
Train	8	13·3	48°
Pedal cycle	10	16·7	60°
Motor cycle	11	18·3	66°
total	60	100	360°

The pie chart is shown in Fig. 7.2.

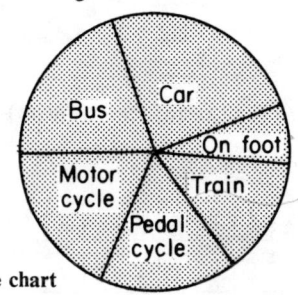

Fig. 7.2 pie chart

EXAMPLE 7.2   A recent survey has shown that, on average, a technician in a particular industry is occupied during an 8 h day, as follows:

Communicating with fellow workers	2 h
Designing	1 h
Installing equipment	4 h
Clerical work	1 h

Display this information pictorially on a pie chart.

The times shown in the table are the number of hours per activity, and can be regarded as frequencies.

Activity	Time (h)	% Time
Communicating	2	25
Designing	1	$12\frac{1}{2}$
Clerical work	1	$12\frac{1}{2}$
Installing	4	50
total	8	100

The pie chart is shown in Fig. 7.3, where it is seen at a glance that the

145

majority of the technician's day is spent communicating and installing equipment.

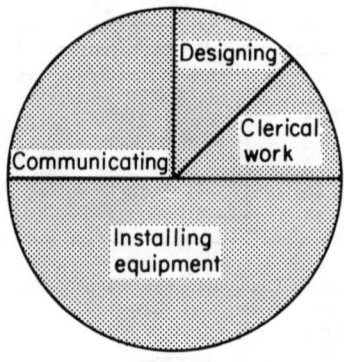

Fig. 7.3

(c) Pictograms

Fig. 7.4 shows a pictogram for the information in Example 7.1.

Mode of travel

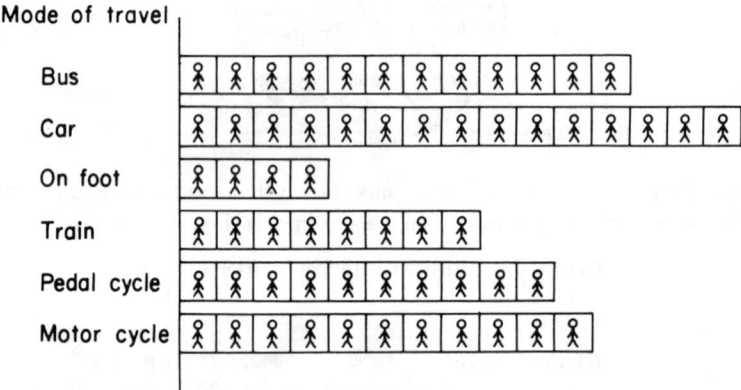

Fig. 7.4 pictogram

Each apprentice is represented by a matchstick figure within a square, that is, 👤

Although pictograms give a good visual effect, the main disadvantage with them is the time required to draw them. It can be seen that the visual effect in Fig. 7.4 is just as good if the matchstick figures are omitted. This principle is used in bar charts described in the next section.

(d) Bar charts

The bars in the bar chart are drawn with their lengths proportional to the frequencies (or relative frequencies). The width of the bar is unimportant. All the bars are drawn at the centre of the class interval.

146

Horizontal or vertical bar charts are equally useful. Both types have been drawn in Example 7.3.

EXAMPLE 7.3   The marks obtained by 180 students in an examination were as follows:

Marks	10–19	20–29	30–39	40–49	50–59	60–69	70–79	80–89
Number of students, i.e., frequency	6	10	16	48	52	28	15	5
Centre of class interval	14·5	24·5	34·5	44·5	54·5	64·5	74·5	84·5

The centre of each class interval has been obtained as follows:

for 10–19 class,   the centre is at $\dfrac{10+19}{2} = 14\!\cdot\!5$

for 20–29 class,   the centre is at $\dfrac{20+29}{2} = 24\!\cdot\!5$

and so on. The bars are now drawn at the positions, 14·5, 24·5, 34·5, ..., 84·5, as shown in Fig. 7.5.

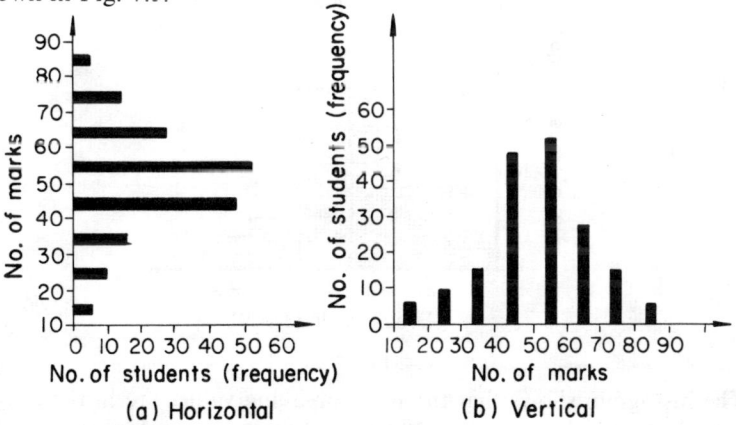

(a) Horizontal   (b) Vertical

Fig. 7.5 bar charts

Note: The frequency scale must always start from zero, otherwise the charts may give a false impression.

From the bar chart it is seen that if 50 is the pass mark, then many students have failed. This could mean that either the students were not properly prepared or the examination was too difficult. A pass mark of 40 only gives 32 failures, which is a more reasonable proportion.

(e) Histogram

The histogram is similar to a vertical bar chart, except that all the bars are widened until each bar or rectangle is equal in width to the class interval. The consecutive rectangles should then be touching each other at the class boundaries, as shown in Example 7.4. If they do not touch at the class boundaries some modifications have to be introduced to make them touch, as shown in Example 7.5.

EXAMPLE 7.4   Fifty steel rods were tested for breaking strength and gave the following results:

Breaking load (kN)	20	21	22	23	24	25
Number of rods	2	8	14	18	5	3

Draw a histogram for these results.

The breaking loads are the centres of the intervals. For example, any breaking load between 19·5 kN and 20·5 kN has been recorded as having a strength of 20 kN. This applies to all the other classes. The histogram is shown in Fig. 7.6.

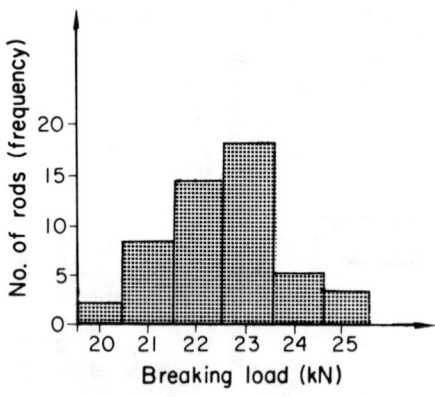

Fig. 7.6

The histogram shows that the main breaking values are in the range 22–23 kN. Few rods fail as low as 20 kN, and few rods reach a value of 25 kN.

EXAMPLE 7.5   Draw a histogram to show the frequency distribution of the information in Example 7.3.

A histogram drawn for the data in Example 7.3 would have its first rectangle stopping at 19 marks, the second at 29 marks, and so on. There would, therefore, be a gap between each rectangle. To prevent this the first two rectangles can be made to touch at 19·5 or, as is more convenient, at the

148

20 mark. The same procedure is then used with all the other rectangles. The histogram is drawn in Fig. 7.7.

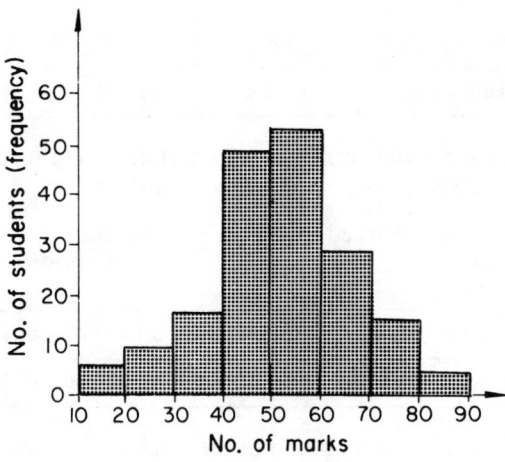

Fig. 7.7 histogram

EXERCISE 7.2

1. In a particular factory the number of different types of technicians employed are

Mechanical	12	Plant	20
Electrical	8	Chemical	45
Building	5		

   Represent these figures pictorially on (a) 100% bar chart, (b) pie chart, (c) pictogram.

2. A survey was carried out by the inspection department in a factory, to determine the pattern of faults in rejected television sets. The results were

Transistor faults	42	Tube faults	14
Assembly faults	86	Resistor faults	18
Condenser faults	32		

   Draw (a) a 100% bar chart, and (b) a pie chart to show the pattern of faults pictorially.

   What should the main recommendations of the department be to reduce faults?

3. Plot bar charts and histograms for the frequency tables obtained in Exercise 7.1.

4. A machine is set to fill packets automatically with 200 g of powder. To check the accuracy of the machine a sample of packets is taken and weighed. The results are shown in the table.

Mass (g)	197	198	199	200	201	202	203	204
Frequency	1	2	2	3	4	12	4	2

Represent these results on (a) a bar chart and (b) a histogram. In your opinion is the machine performing satisfactorily?

5. The overtime hours worked in a particular factory for a typical month were as follows:

Number of hours	32	34	36	38	40	42	44	46
Number of employees	10	12	14	18	12	40	16	5

Draw a pictorial representation of these results, and hence describe the pattern of overtime worked in the factory.

6. The lifetime of transistors from a production line was tested with a batch of 80 transistors. The results obtained are shown in the table.

Lifetime in hours	600–619	620–639	640–659	660–679	680–699	700–719	720–739	740–759	760–779	780–799
Frequency	1	2	3	7	10	13	20	18	4	2

Draw a histogram to represent the frequency distribution. How would you describe to a new customer the length of time a transistor will operate before failing?

7. A machine is set to pack paper clips in boxes of 100. As a result of complaints it was decided to test the performance of the machine by counting the number of clips in a sample of boxes. The results are shown in the table.

Number of clips in the box	84	88	92	96	100	104	108	112
Frequency	21	25	26	28	30	10	6	2

Draw a horizontal bar chart and a histogram from the table.

Do the complaints seem justified, and if so, what action should be taken?

Assessment test 7

1. Select the correct name from the list to match each of the diagrams in the figure, by filling in the appropriate number in each box.

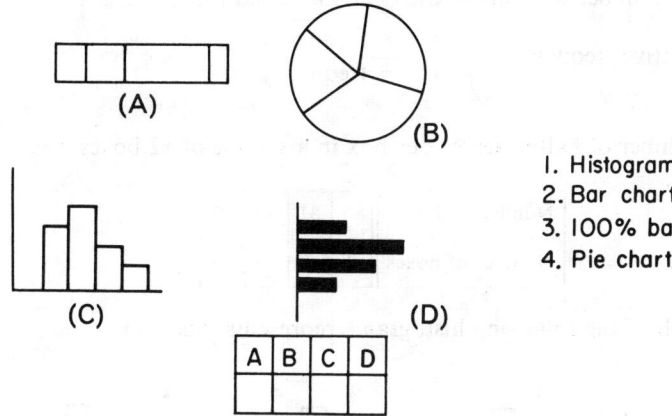

1. Histogram
2. Bar chart
3. 100% bar chart
4. Pie chart

Fig. AT 7.1

2. Which of the following 100% bar charts represents correctly the data in the table.

Length of rod (m)	3·0	3·1	3·2	3·3
Relative frequency (%)	10	40	20	30

(a)   (b)   (c)   (d)

3. Complete the following statements,
   (a) Data is usually grouped in ....
   (b) The spread of data from the smallest to the largest values is called the ....
   (c) The number of items in each class is called the ....
   (d) Relative frequency $= \dfrac{*\ *\ *\ *\ \text{frequency}}{*\ *\ *\ *\ \text{frequency}}$

4. The number of ball bearings per box in a sample of 12 boxes was

Number in box	80	81	82	83
Number of boxes	1	2	6	3

   Which of the following histograms represents this data?

   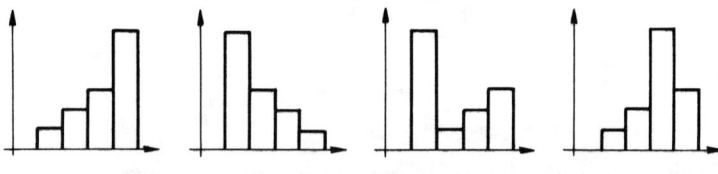

   Fig. AT 7.2

5. Which of the following pie charts is a true representation of the data in question 4.

   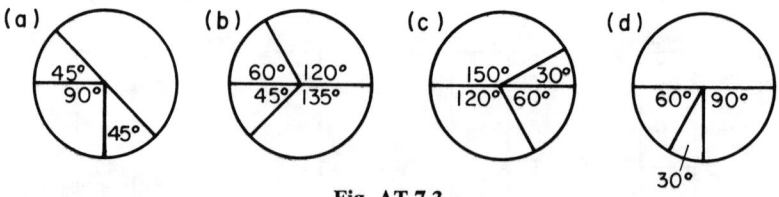

   Fig. AT 7.3

6. Complete the frequency table giving the diameters of a sample of steel bars.

Diameter of bar (cm)	0·875–0·915	0·915–0·955			
Relative frequency (%)		18	32	15	5

7. Which of the following 100% bar charts is **not** correct?

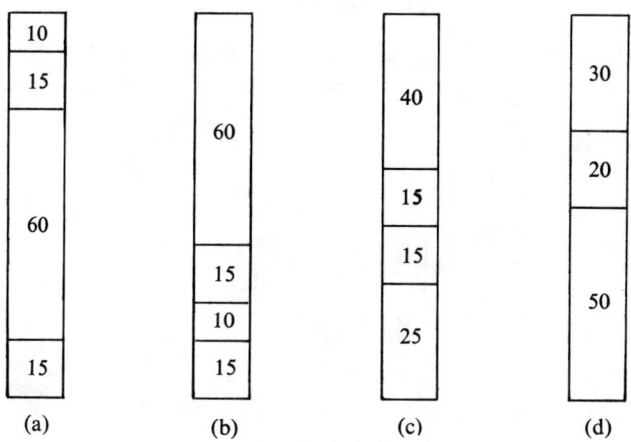

Fig. AT 7.4

8. To draw a bar chart or histogram the number of classes should be, if possible,
   (a) an even number
   (b) an odd number
   (c) less than 5
   (d) 5–15
   Choose **one** correct answer.

9. A certain class interval is 30 39·5. Which of the following is the centre of this class interval?

   (a) 35
   (b) 34·75
   (c) 35·5
   (d) 34·5

10. State whether the following statements are **true** or **false**.
    (a) The range is $\dfrac{\text{highest class value} + \text{lowest class value}}{2}$

    (b) Number of classes is $\dfrac{\text{range}}{\text{class interval}}$

    (c) The angle between the radii of a sector in a pie chart is proportional to the class frequency.

(d) Relative frequency is $\dfrac{\text{total frequency}}{\text{class frequency}}$

11. The histogram in the diagram has been drawn to scale, but the third rectangle C has been omitted. Complete the histogram by drawing in this rectangle C.

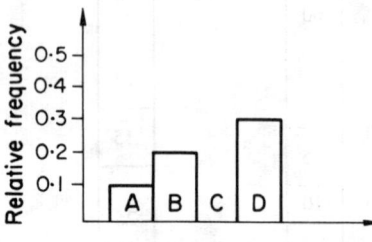

Fig. AT 7.5

12. The relative frequency of a particular class is $\frac{1}{4}$. When expressed as a percentage the class has a relative frequency of

(a) $\frac{1}{4}\%$

(b) 25%

(c) 40%

(d) 4%

Choose **one** correct answer.

13. Four samples of resistors are taken from a machine producing resistors with a nominal (i.e., desired) value of 100 ohm. A histogram is drawn for each sample shown in the figure. Which sample indicates the most satisfactory production process?

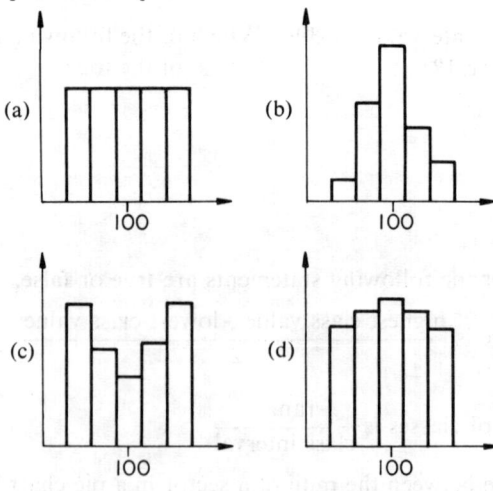

Fig. AT 7.6

154

8. Geometry 1

Objectives

After working through this chapter you should be able to

1. Explain the terms angle, degree, minute.
2. Identify acute, right, obtuse, straight and reflex angles.
3. State the properties of complementary, supplementary, and vertically opposite angles.
4. Calculate unknown angles using the properties listed in 3.
5. Explain the terms parallel line and transverse.
6. State the properties of alternate, corresponding, and interior angles.
7. Calculate for given parallel lines and a transverse unknown angles using the angle properties listed in 6.
8. Prove that two lines are parallel by using the angle properties listed in 3 and 6.
9. State the sum of the angles in a triangle.
10. Use the 'sum of the angles of a triangle' to calculate unknown angles in a triangle.
11. List the properties of actue-angled, right-angled, obtuse-angled, equilateral, and isosceles triangles in relation to their sides and angles.
12. Determine unknown angles in the triangles listed in 11.
13. State Pythagoras' theorem.
14. Use Pythagoras' theorem to calculate any third side of a right-angled triangle.
15. Construct a right-angled triangle by drawing a triangle with sides in the ratio $3 : 4 : 5$.
16. List the conditions of congruency of two triangles.
17. State the conditions of similarity for two triangles.
18. Use the conditions of congruency and similarity to determine unknown sides and angles in a second triangle.
19. Construct a triangle given one of the following sets of information.
    (a) three sides
    (b) two sides and an included angle
    (c) one side and two angles
    (d) the hypotenuse, another side, and a right angle.

Geometry is a branch of mathematics that deals with angles and lines and the relationships that exist between them.

8.1 Angles

An **angle** is the amount of rotation between two straight lines. Consider a line OX fixed in position and another line OP free to rotate about point O.

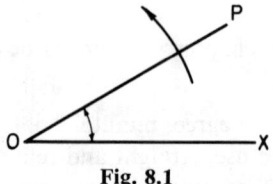

Fig. 8.1

In Fig. 8.1, OP has rotated from OX. The amount of rotation is known as the angle POX (written ∠POX). Angles are normally measured in degrees, where 360 degrees (written 360°) represents one complete revolution, i.e., when OP has rotated all the way around until it reaches OX again.

A degree is subdivided into 60 minutes (written 60′)

Angles are named according to their size as listed below with illustrations:

(a) acute angle, any angle between 0° and 90°;

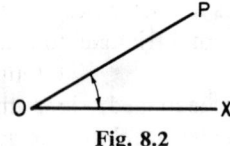

Fig. 8.2

(b) right angle, an angle equal to 90°;

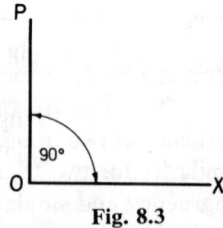

Fig. 8.3

(c) obtuse angle, any angle between 90° and 180°;

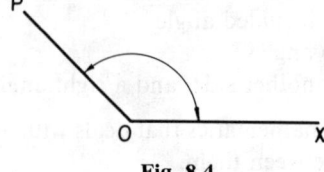

Fig. 8.4

(d) straight angle, an angle equal to 180°;

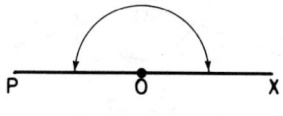

Fig. 8.5

(e) reflex angle, any angle greater than 180°.

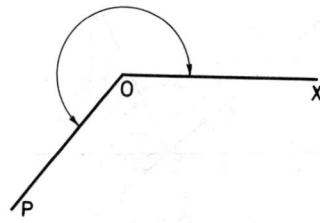

Fig. 8.6

Note: In case (b) the lines PO and OX are said to be perpendicular to one another or normal to each other. In case (d) POX is a straight line.

8.2 Properties of angles

Complementary angles

If two angles add up to 90°, they are called complementary angles, e.g., 60° and 30° are complementary angles.

Supplementary angles

If two angles add up to 180°, they are called supplementary angles, e.g., 45° is the supplementary angle to 135°.

Vertically opposite angles

Consider two straight lines AB and CD meeting at the point O (Fig. 8.7). For simplicity let $\angle$AOD be called α, $\angle$DOB be called β, etc. Angles α and α_1 are vertically opposite each other, and it is obvious that they are equal, that is,

$$\alpha = \alpha_1$$
$$\beta = \beta_1$$

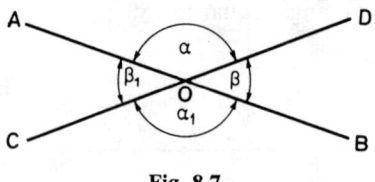

Fig. 8.7

EXAMPLE 8.1 AOB and COD are two intersecting straight lines and E is a point between B and C. If $\angle BOD = 40°$ and $\angle EOB = 65°$ find the unknown angles α, β and θ.

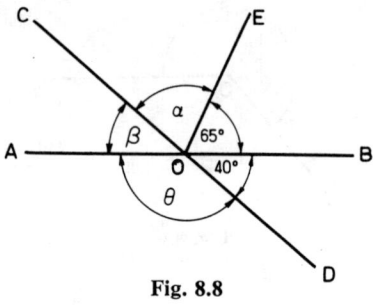

Fig. 8.8

In Fig. 8.8 since COD is a straight line, then by supplementary angles

$$\alpha = 180 - (40 + 65)$$

$$= 75°$$

Now angles COA and BOD are vertically opposite

$$\beta = 40°$$

Since AOB is a straight line, then by supplementary angles

$$\theta = 180 - 40$$

$$= 140°$$

Angles made with parallel lines

Note: **Parallel** lines are lines which never meet, however far they are produced in either direction. Therefore, since they never meet, there can be no angle between them.

Also a **transversal** is a line cutting across parallel lines.

In Fig. 8.9 below, AB is parallel to CD (written AB∥CD) and GH is the transversal.

Parallel lines are very often identified by arrowheads.

158

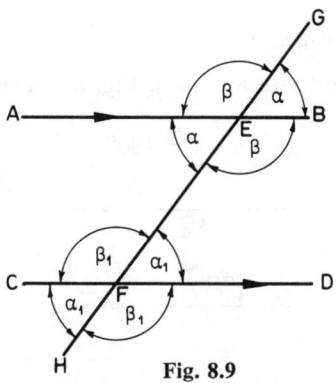

Fig. 8.9

The vertically opposite angles at each point of intersection have been marked. Since the lines are parallel then GH is equally inclined to both lines, i.e., $\angle GEB = \angle EFD$ and $\angle FEB = \angle HFD$, or $\alpha = \alpha_1$ and $\beta = \beta_1$. The results can be summarized as follows.

Alternate angles are equal. These are marked with the same letter in Fig. 8.10.

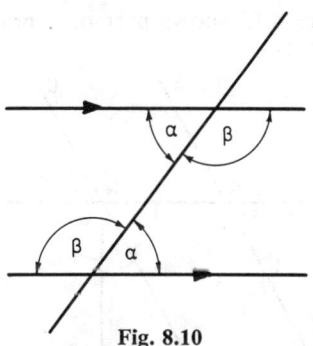

Fig. 8.10

Corresponding angles are equal. There are four pairs of corresponding angles in Fig. 8.9. Two pairs are marked in Fig. 8.11.

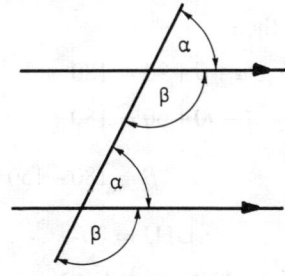

Fig. 8.11

159

Interior angles are supplementary

There are two pairs of interior angles in Fig. 8.9. These are shown again in Fig. 8.12, where

$$\alpha + \beta = 180°$$

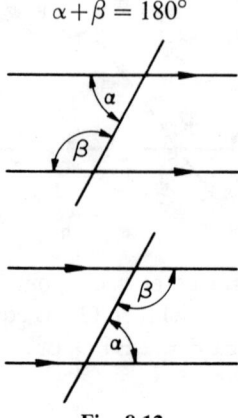

Fig. 8.12

EXAMPLE 8.2   Figure 8.13 shows part of a bridge frame where AB is parallel to CD. Find $\angle DHJ$

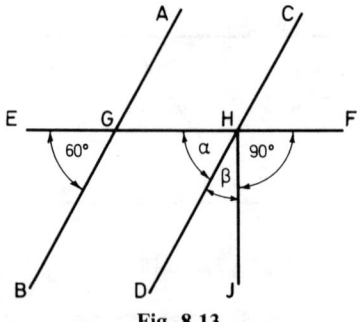

Fig. 8.13

$$\alpha = \angle EGB = 60° \text{ (corresponding angles)}$$

Since EF is a straight line, then

$$\alpha + \beta + 90 = 180$$
$$\beta + 90 + 60 = 180$$

Therefore

$$\beta = 180 - 150$$

that is

$$\angle DHJ = 30°$$

These results may also be used to prove that two lines are parallel as shown in Example 8.3.

160

EXAMPLE 8.3   ABCD in Fig. 8.14 is the plan view of a construction site. Show that the boundaries AB and CD are parallel to each other.

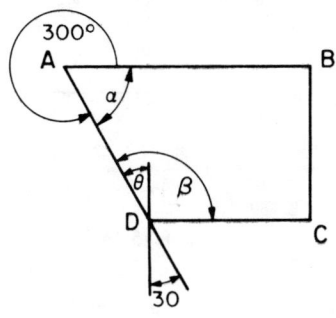

Fig. 8.14

$$\alpha = 360 - 300 = \quad 60° \quad \text{(one revolution)}$$
$$\theta = \quad 30° \qquad \qquad \text{(vertically opposite)}$$

Therefore

$$\beta = \quad 90 + 30 = 120°$$

Hence

$$\alpha + \beta = \quad 60 + 120 = 180°$$

Thus AB∥CD because the interior angles add up to 180°.

EXERCISE 8.1

1. Determine the unknown angles α, β, and γ in Fig. 8.15.

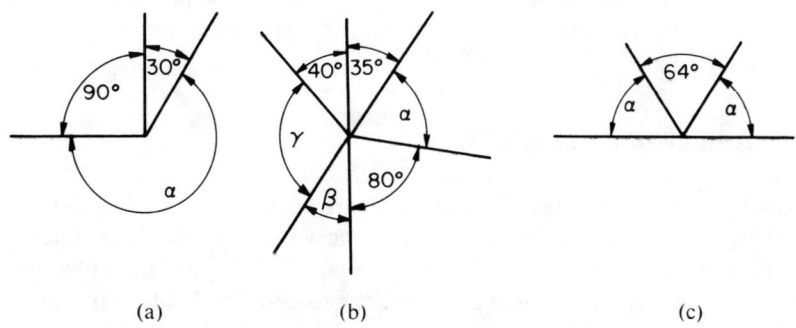

(a)   (b)   (c)

Fig. 8.15

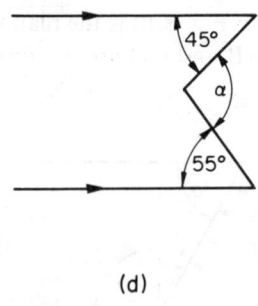

(d)

Fig. 8.15 (*cont.*)

2. (a) Find the angles complementary to (i) 49°, (ii) 37° 21′, (iii) 79° 6′.
   (b) Find the angles supplementary to (i) 78°, (ii) 80° 12′, (iii) 147° 49′.
3. If angle AOB is 29°, find the reflex angle AOB.
4. AOD is a straight line. If B and C are two points on the same side of this line such that ∠BOC = 90° and ∠BOA = 30° determine the value of ∠COD.
5. AB and BC are two links perpendicular to one another. D is a point on AC such that ∠ABD = 72°. What is the angle between links BD and BC?
6. AOB is a straight line and line CO is drawn such that angle COA = 55°. If the obtuse angle COB is bisected by a line DO, find ∠DOB.
7. EF is a line cutting two parallel lines AB and CD at points G and H respectively. If ∠EGB = 65° find all the unknown angles.
8. Two straight roads AB and CD are intersected by another road EF at points G and H respectively. If angle EGB = 70° and angle CHG = 110°, prove that the roads AB and CD are parallel.
9. A gear wheel has 40 teeth. Find the angle of rotation of the wheel when (a) four teeth and (b) 25 teeth pass a given mark.
10. ABCD is a frame with AB parallel to DC, AD parallel to BC and ∠ADC = 58°. What are the remaining angles of the frame?

8.3   Angles of a triangle

Definition: *A triangle is a closed plane figure bounded by three straight lines.*
   The symbol '△' is often used instead of the word 'triangle'. These lines meet at points which are labelled with capital letters. Such points are known as the **vertices** of the triangle. There are three angles contained within the triangle.
   Consider triangle ABC with side BC produced to D and CE drawn parallel to AB (Fig. 8.16).

162

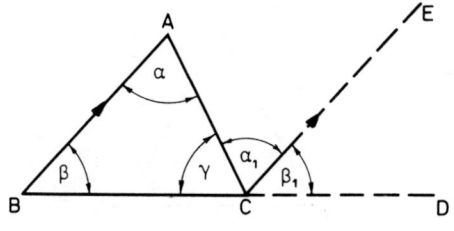

Fig. 8.16

Since AB and CE are parallel then

$$\alpha = \alpha_1 \quad \text{(alternate angles)}$$

also

$$\beta = \beta_1 \quad \text{(corresponding angles)}$$

Since BCD is a straight line,

$$\alpha_1 + \beta_1 + \gamma = 180°$$

or

$$\alpha + \beta + \gamma = 180°$$

The sum of the angles in a triangle is 180°.

Triangles are named according to their angles:
(a) **Scalene** triangle—a triangle with unequal angles and sides.
(b) **Acute-angled** triangle—a triangle with all angles acute.
(c) **Obtuse-angled** triangle—a triangle with one obtuse angle. (Since the angles in a triangle add up to 180° it cannot have two angles greater than 90°, i.e., it cannot have two obtuse angles.)
(d) **Right-angled** triangle—a triangle containing a right angle. The side opposite this right angle is called the 'hypotenuse'.
    Since one angle is 90° then the other two angles in a right-angled triangle must add up to 90°, that is they are complementary angles. In Fig. 8.17:

$$\alpha + \beta = 90°$$

Fig. 8.17

(e) **Isosceles triangle**—a triangle containing two equal angles; the sides opposite these angles are also equal and this is shown in Fig. 8.18, where $\angle B = \angle C$ and AB = AC.

163

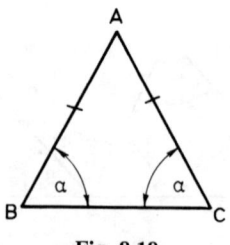

Fig. 8.18

(f) **Equilateral** triangle—a triangle with all its sides equal (since the sum of the angles = 180° then each angle must be 60°).

EXAMPLE 8.4   Two rods AB and AC of equal length are welded to rod BD as shown in Fig. 8.19. If the angle A is 48° what is angle β?

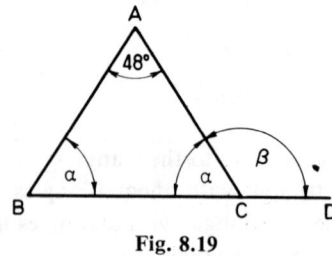

Fig. 8.19

In Fig. 8.19, since the angles in $\triangle ABC$ add up to 180°, then

$$\alpha + \alpha + 48 = 180$$

$$2\alpha = 180 - 48$$

$$2\alpha = 132$$

$$\alpha = 66°$$

Since BCD is a straight line, then

$$\alpha + \beta = 180$$

Therefore

$$66 + \beta = 180$$

$$\beta = 180 - 66$$

$$= 114°$$

EXERCISE 8.2

1. Find the unknown angles α and β in Fig. 8.20.

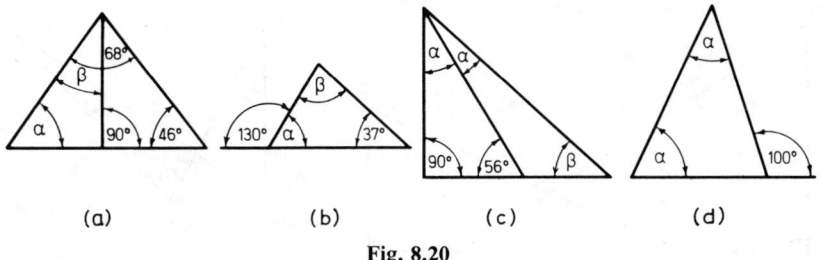

(a)   (b)   (c)   (d)

Fig. 8.20

2. In a right-angled triangle one angle is 49° 12′, find the value of the other acute angle.
3. In △ABC, ∠A = 27° and ∠C = 92°. Find the remaining angle. What sort of triangle is this?
4. The end view of a roof-top is triangular in which the angle at the apex is 120°. What angle does each side make with the horizontal, if the sides are equal?
5. A triangular gauge ABC has its angle A = 58°. When the gauge is made to stand upright with BC on a horizontal surface the side AB is at 34° to the vertical. Find the angles of the gauge.
6. A wall crane consists of a tie AB and a strut CB, hinged to a vertical wall at A and C. If the tie is at 45° to the wall and the strut at 30° to the wall, find the angle between the tie and the strut.
7. Determine the value of the angles α, β, and θ in Fig. 8.21. Parallel lines are marked with arrowheads.

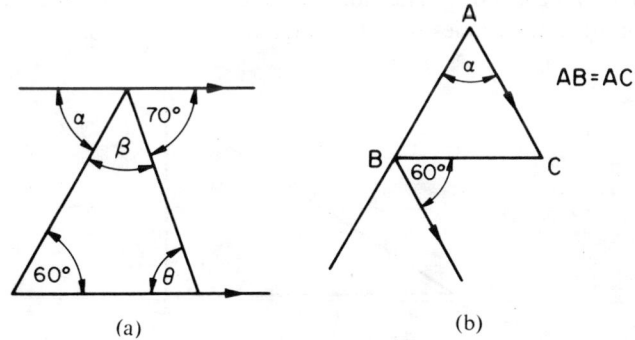

(a)   (b)

Fig. 8.21

8. In Fig. 8.22(a) calculate the angles DAE, CBE, and BEC if angle BDC = 35° and angle ACD = 42°.

165

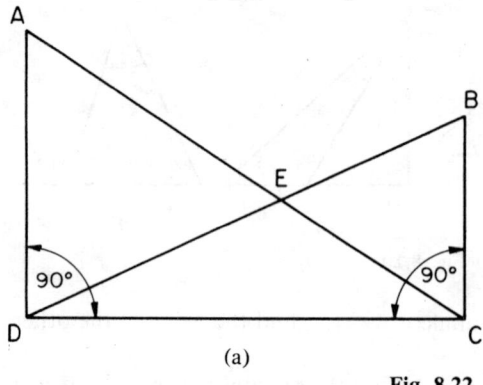

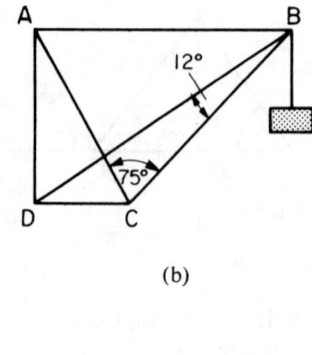

(b)

(a)

Fig. 8.22

9. A frame is to be made to support a load at B (Fig. 8.22b). If the girder BC must be inclined at 45° to the horizontal find the inclination of BD to the horizontal.

10. A vertical radio mast AB rests on its base. A staywire AD is fixed to the ground at D and makes an angle of 40° with the mast. Another staywire CD from a point C on the mast makes an angle of 65° with the mast. What is the angle between the staywires?

8.4 Pythagoras' theorem

Theorem. *For any right-angled triangle the square of the hypotenuse is equal to the sum of the squares on the other two sides.*

Consider triangle ABC. The side opposite angle A is labelled a, etc. (this method of labelling applies to any triangle).

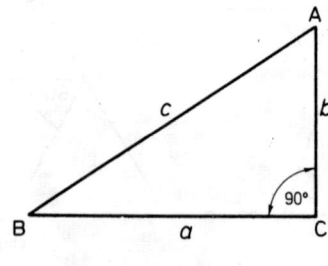

Fig. 8.23

The theorem states that

$$(AB)^2 = (BC)^2 + (AC)^2$$
$$c^2 = a^2 + b^2$$

166

The theorem may be proved quite easily using Fig. 8.24.

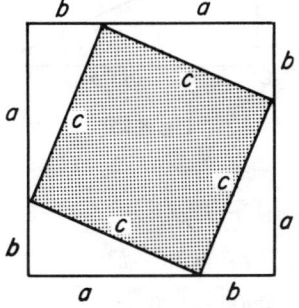

Fig. 8.24

The larger square contains a smaller inner square. The four triangles so formed are identical. They have sides a, b, and c. Now

area of small square + area of four triangles = area of large square

$$c^2 + 4 \times \tfrac{1}{2}ab = (a+b)^2$$

$$c^2 + 2ab = a^2 + 2ab + b^2$$

$$c^2 = a^2 + b^2$$

There is an infinite (unlimited) number of combinations of sides which make a right-angled triangle, but it is useful to bear in mind two particular triangles:

(a) sides in ratio 3 : 4 : 5   (b) sides in ratio 5 : 12 : 13

$$c^2 = a^2 + b^2 \qquad\qquad c^2 = a^2 + b^2$$

$$= 3^2 + 4^2 \qquad\qquad\quad = 5^2 + 12^2$$

$$= 9 + 16 \qquad\qquad\quad = 25 + 144$$

$$= 25 \qquad\qquad\qquad\ = 169$$

$$c = \sqrt{25} \qquad\qquad\quad c = \sqrt{169}$$

$$= 5 \qquad\qquad\qquad\ = 13$$

This theorem is useful for calculating the length of any side of a right-angled triangle, given the lengths of the other two sides, as shown in Example 8.5.

EXAMPLE 8.5   Calculate the length of the side of a right-angled triangle if one side is 30 mm long and the hypotenuse is 70 mm long (Fig. 8.25).

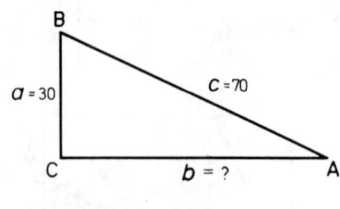

Fig. 8.25

By Pythagoras

$$c^2 = a^2 + b^2$$

$$70^2 = 30^2 + b^2$$

$$4900 = 900 + b^2$$

$$\therefore b^2 = 4900 - 900 = 4000$$

$$b = \sqrt{4000}$$

$$= 63 \cdot 3 \text{ mm}$$

EXERCISE 8.3

1. Triangle ABC is a right-angled triangle with $\angle$ B = 90°.
   (a) If $a = 10$ and $c = 14$ find b.
   (b) If $a = 3$ and $b = 6$ find c.
2. Find the length of the diagonal (i.e., a line joining opposite corners) of a square of side 100 mm.
3. A rectangle has a width of 82 mm and a diagonal of length 126 mm. Find the length of the rectangle.
4. What is the length of the side of a square socket which has a diagonal 16 mm long?
5. A ladder of length 22·5 m leans against a vertical wall and touches it at a point 19·0 m from the ground. How far is the foot of the ladder from the wall?
6. The feet of two vertical posts are 30 m apart on level ground. If the posts are 6 m and 10 m high respectively, what is the minimum length of cable required to join the tops of the two posts?
7. Calculate the longest thin straight wire that can be placed in a closed rectangular box $50 \times 120 \times 150$ mm.

8.5 Construction of a right angle

In Section 8.4 it was shown that a triangle with sides of the ratio 3 : 4 : 5 is right-angled. This information can be used to construct an accurate right angle, using a compass and rule only. The method is illustrated in Fig. 8.26.

168

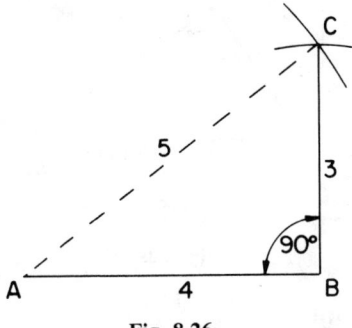

Fig. 8.26

(a) Draw AB 4 units long.

(b) Set the compass to 5 units long, and with the compass point at A draw an arc above B.

(c) Set the compass to 3 units, and with the compass point at B draw an arc to cut the first arc at C.

(d) Join B to C.

The angle ABC is now 90°.

Note: An accurate angle will not be obtained if the figure drawn is too small.

EXERCISE 8.4

1. Three holes are to be marked on a television chassis, as shown in Fig. 8.27

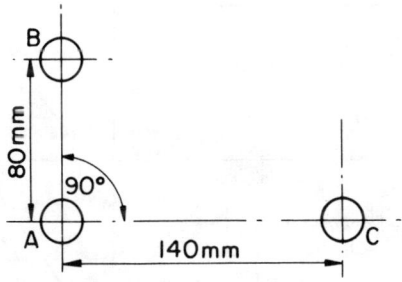

Fig. 8.27

It is important that the holes are precisely located. To check the accuracy the distance BC is measured.

Draw the figure to scale in order to determine the exact length of BC. Use a calculation as a check.

2. A roof truss for a factory is required to have the conditions specified in Fig. 8.28. Draw the truss to scale, constructing the right angles using a compass and rule only. From the drawing determine

(a) the roof span s

(b) the length of steel required to make the triangular truss.

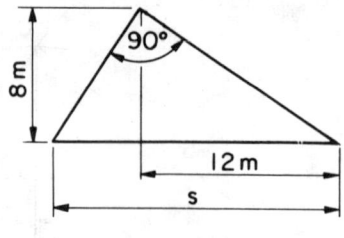

Fig. 8.28

8.6 Congruent triangles

Triangles are said to be **congruent** if they are exactly the same, i.e., three sides and three angles in one triangle are equal to three sides and three angles in the other triangle.

Conditions of congruency

There are six possible facts to be known about a triangle, namely three sides and three angles. However, to show that two triangles are congruent it is only necessary to show that three facts are equal in both triangles—provided they are the correct three facts. Two triangles are congruent if the following groups of three facts are equal in both triangles:

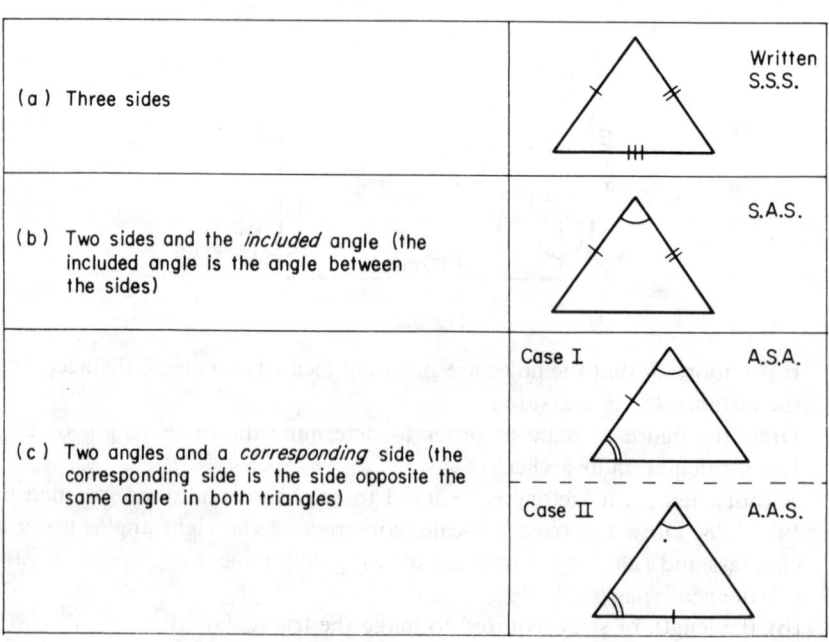

(d) A right angle, hypotenuse, and any other side	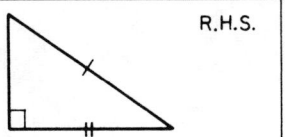 R.H.S.

EXAMPLE 8.6 Two equal rods AB and AC are standing on a horizontal floor as shown in Fig. 8.29. Prove that the plumb-line hanging freely from A will locate the point D on the floor midway between B and C

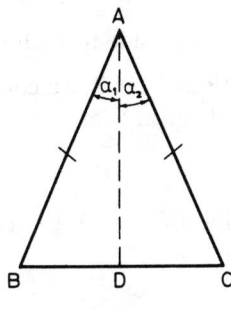

Fig. 8.29

Compare triangles ABD and ACD.

$\angle$ ADB = $\angle$ ADC   (AD vertical, so both angles are 90°)

AB = AC   (length of the two rods)

AD = is a common side

Therefore the triangles ABD and ACD are congruent (R.H.S.). Hence

$$BD = DC$$

which means that D is the mid-point of BC.

Note also that $\angle$ A is bisected since $\alpha_1 = \alpha_2$.

EXAMPLE 8.7 Prove that any four-sided flat sheet of metal with its opposite sides equal, can be cut into two identical triangular parts by cutting along a diagonal. The sheet of metal is shown in Fig. 8.30.

171

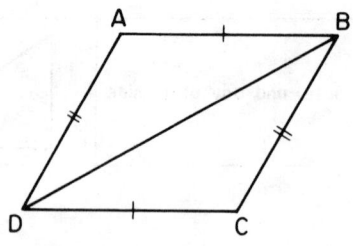

Fig. 8.30

Compare triangles ADB and DCB in the figure. Now

AB = CD   (given condition)

AD = BC   (given condition)

BD is a common side to both triangles.

Therefore the triangles ADB and DCB are congruent (S.S.S.), that is, the diagonal cuts the figure into identical triangles.

EXERCISE 8.5

1. In Fig. 8.31(a) AB = AD and BC = CD. Use congruent triangles to prove that ABCD is symmetrical.

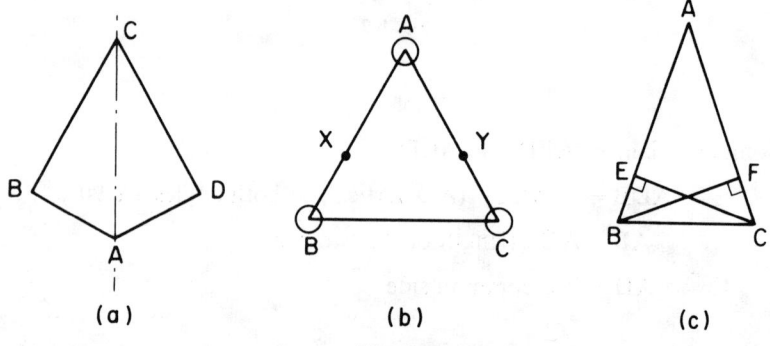

(a)   (b)   (c)

Fig. 8.31

2. ABCD is a rectangular gauge. Perpendiculars are drawn to the diagonal BD from points A and C and meet it at E and F respectively. By means of congruent triangles prove that the holes drilled at E and F are the same distance from the corners B and D respectively.

3. Three girders form an isosceles triangle ABC, with AB = AC. Points P and Q are now chosen on the girder BC, such that BP = QC. Prove that if two extra girders have to be placed along AP and AQ, they will be of the same length.

172

4. The pointer AB of an instrument is pivoted at its centre. Use congruent triangles to prove that if the end A moves through a certain distance, the end B moves through the same distance.

5. Three holes A, B, and C are drilled at the corners of an equilateral triangle (Fig. 8.31b). Two transistors are now placed at X and Y, such that AX = AY. Show that the distances XC and YB are equal.

6. A triangular frame ABC is in the form of an isosceles triangle with AB = AC (Fig. 8.31c). Extra rods BF and EC are now used to support the sides AB and AC. Prove that these rods must be of equal length.

7. In triangle ABC, AB = AC and X and Y are points on AB and AC respectively, such that AX = AY. Show that the triangles AYB and AXC are congruent.

8. Triangle ABC has all its internal angles bisected by lines which meet the opposite sides at 90°. What type of triangle is this?

8.7 Similar triangles

Two triangles are said to be **similar** if the angles in one triangle are equal to the angles in the other triangle. The corresponding sides in both triangles will not necessarily be equal but will be in proportion, i.e., one triangle is an enlargement of the other.

Consider the two similar triangles ABC and PQR (Fig. 8.32).

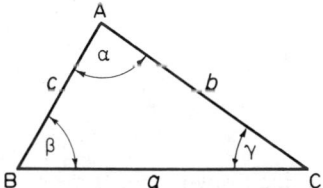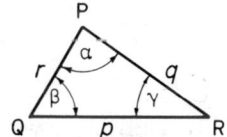

Fig. 8.32

Since the sides are in proportion

$$\frac{a}{p} = \frac{b}{q} = \frac{c}{r}$$

EXAMPLE 8.8 It is necessary to strengthen a triangular crane (Fig. 8.33) by including an extra rod DE. Use similar triangles to find the length of this rod.

Compare triangles ADE and ABC in Fig. 8.33.

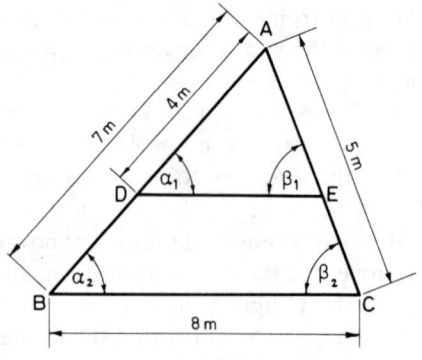

Fig. 8.33

Since DE is parallel to BC

$$\alpha_1 = \alpha_2 \quad \text{(corresponding angles)}$$

$$\beta_1 = \beta_2 \quad \text{(corresponding angles)}$$

also $\angle$ A is common to both triangles. Since the angles in both triangles are equal, the triangles are similar. Hence the sides are in proportion, i.e.,

$$\frac{DE}{BC} = \frac{AD}{AB}$$

$$DE = \frac{8 \times 4}{7}$$

$$= 4{\cdot}57$$

$$= 4{\cdot}6 \text{ m}$$

EXERCISE 8.6

1. ABC is a triangle with $\angle$ ABC $= 90°$. A line drawn from B meets AC at right angles at point D. Show that all the triangles are similar.
2. Triangle ABC has DE drawn parallel to BC where D and E are points on AB and AC respectively. If AD $= 9$, DB $= 2$, and AC $= 13$ find the length of AE.
3. ABCD is a flat plate with AD parallel to BC. If the diagonals meet at X show that triangles AXD and BXC are similar.
4. Find the height of a tower which throws a shadow of 30 m if a man of height 1·82 m throws a shadow of 3·2 m. Give your answer correct to three significant figures.
5. ABC is a triangle, D and E are points on AB and AC such that DE is parallel to BC and is 7 mm from it. Find the vertical height of triangle ADE if DE $= 5$ mm and BC $= 8$ mm.

6. A frame ABCD has AB parallel to CD. The diagonals meet at point X. If AB = 5 m, DC = 7 m, AX = 6 m, and BX = 1 m, find the lengths of DX and CX using similar triangles.

7. Find the lengths of the unknown sides in the triangular template DEF (Fig. 8.34).

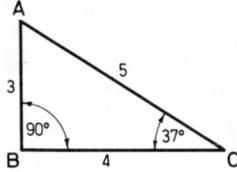 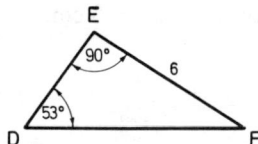

Fig. 8.34

8. Triangle ABC has ∠ B = 90°. D is a point on AC, E is a point on AB, and F is a point on BC. If DE is perpendicular to AB and DF is perpendicular to BC, BC = 8 mm, DF = 4 mm, and AE = 3 mm, find the length of BF.

9. Figure 8.35 shows a ladder passing over a lean-to garage 2·7 m wide by 2·4 m high. If the foot of the ladder is 1·0 m from the garage what is the height reached up the wall?

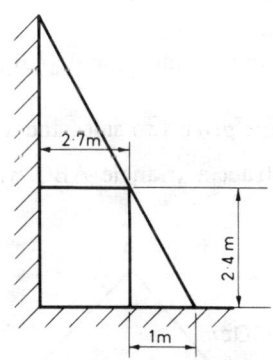

Fig. 8.35

10. Calculate the height of a post 25 m away from a man if a rod 300 mm long held 650 mm away from his eye just hides the post from view. Assume the man is lying on the floor making this observation.

11. A pair of steps has a vertical height of 1·520 m when the feet are 1·20 m apart. Calculate the length of horizontal rope required, 600 mm from the ground to hold the two parts of the steps together.

8.8 Construction of triangles

Triangles may be constructed provided at least three suitable facts are known. The different combinations of the suitable facts are described below.

(a) Construction of a triangle given the lengths of three sides

EXAMPLE 8.9 Construct a triangle ABC with sides AB = 4 cm, BC = 6 cm, and AC = 8 cm.

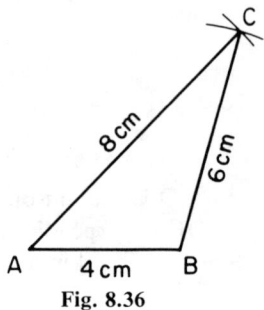

Fig. 8.36

(a) Draw AB = 4 cm, as in Fig. 8.36.
(b) Set a compass to 8 cm, and with the compass point at A draw an arc above AB.
(c) Set the compass to 6 cm, and with the compass point at B draw an arc to cut the previous one at C.
(d) Join AC and BC to complete the triangle ABC.

(b) Construction of a triangle given two sides and an included angle

EXAMPLE 8.10 Construct a triangle ABC with $b - 14$ cm, $c = 10$ cm, and A = 50°.

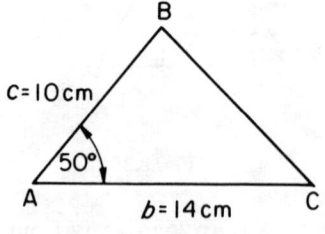

Fig. 8.37

(a) Draw a line AC = 14 cm (Fig. 8.37).
(b) With a protractor at A set up a line through A at 50° to AC.
(c) Mark a point B along this line, such that AB = c = 10 cm.
(d) Join BC to complete the triangle ABC.

(c) Construction of a triangle given one side and two angles

It is sufficient to know any two angles since the third can easily be found if required.

EXAMPLE 8.11 Draw a triangle ABC given that B = 80°, C = 40°, and b = 700 mm.

Now the angles of a triangle add up to 180°. Therefore

$$\angle A + \angle B + \angle C = 180°$$

$$\angle A + 80 + 40 = 180°$$

$$\angle A = 60°$$

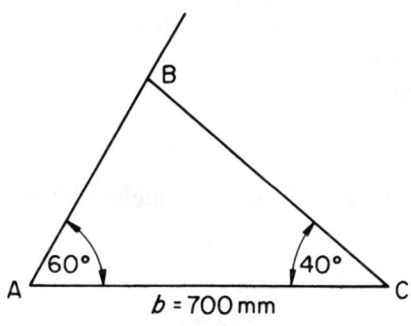

Fig. 8.38

The two angles ∠ A and ∠ C, which are at the ends of the given line, are used to construct the triangle, as shown in Fig. 8.38.

(a) Draw AC = b = 700 mm to a suitable scale.

(b) Use a protractor to draw a line through A at 60° to AC.

(c) Use a protractor to draw a line through C at 40° to AC.

(d) Let the two lines so drawn cut each other at B.

Complete the triangle ABC.

(d) Construction of a triangle, given the hypotenuse, a right angle, and another side

EXAMPLE 8.12 Draw a triangle ABC, with ∠ C = 90°, AC = 800 mm, and AB = 930 mm.

Since AB is opposite the right angle C, then it must be the hypotenuse of the triangle.

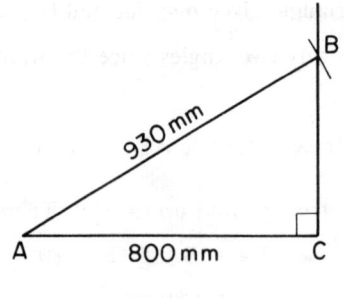

Fig. 8.39

(a) Draw AC = 800 mm to a suitable scale (Fig. 8.39).

(b) Draw a line perpendicular to AC, using a protractor, or the compass method given in Section 8.5.

(c) Set the compass to 930 mm, and with the compass point at A draw an arc to cut the perpendicular to AC at B.

(d) Complete the triangle ABC.

EXERCISE 8.7

1. Draw, to scale, the plan view of a triangular plot of land ABC given that the sides of the plot are $a = 6$ km, $b = 7$ km, and $c = 10$ km.

2. Draw to scale a triangle which has two sides 100 mm long, with an angle of 60° between them. What type of triangle have you constructed?

3. It is required to produce a triangular steel plate, which has one edge 1·2 m long, with the other two edges inclined at 100° and 30° to it. Draw a scale plan of the required plate.

4. A support bracket is to be made in the form of a right-angled triangle by welding three steel rods together. Two rods available are 400 mm and 1200 mm long. Find by drawing the bracket to scale the length of the third rod required. There is a condition that no rod longer than 1200 mm is available.

Assessment test 8

1. Complete the following statements:

    (a) An angle is the amount of .... between two lines.

    (b) .... degrees represents one complete revolution.

    (c) There are .... minutes in one degree.

    (d) The sum of the angles of a triangle is ....

2. List I gives diagrams of four different angles. List II gives their names. Match the correct name to the angle by filling in the appropriate number in the boxes.

List I

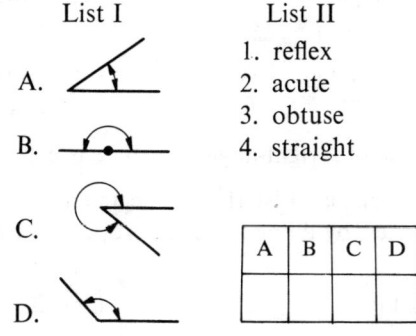

List II
1. reflex
2. acute
3. obtuse
4. straight

A	B	C	D

Fig. AT 8.1

3. The two figures show a pair of angles α and β. Which of the following statements is true for these two angles in **both** figures?

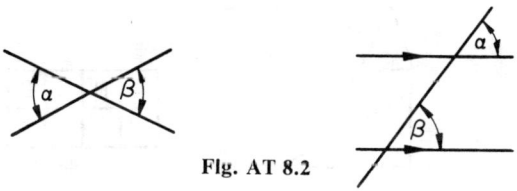

Fig. AT 8.2

Angles α and β are
(a) vertically opposite (c) corresponding
(b) equal   (d) supplementary

4. Which of the following describes the pair of angles α and β?

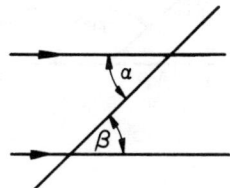

(a) not equal
(b) equal
(c) acute
(d) supplementary
(e) alternate

Fig. AT 8.3

179

5. Complete the following statements:

(a) A.scalene triangle has .... sides equal and .... angles equal.

(b) An isosceles triangle has .... sides equal and .... angles equal.

(c) An equilateral triangle has .... sides equal and .... angles equal.

(d) The longest side of a right-angled triangle is called the ....

6. List I gives four triangles. List II gives their names. Match the correct name to the triangles, by filling in the appropriate number in the boxes.

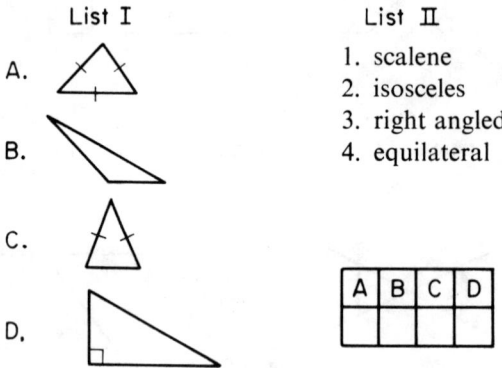

List I

A.

B.

C.

D.

List II

1. scalene
2. isosceles
3. right angled
4. equilateral

A	B	C	D

Fig. AT 8.4

7. Select the correct value of α from the following:

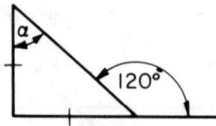

(a)   30°
(b)   60°
(c)   90°
(d) 120°

Fig. AT 8.5

8. In the diagram a number of angles are labelled.

180

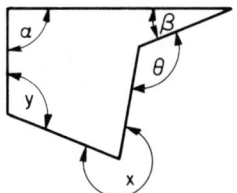

Fig. AT 8.6

Write in the correct angle in each of the blank spaces below.

(a) the reflex angle is ....

(b) the acute angle is ....

(c) the obtuse angle is ....

(d) the right angle is ....

9. In the diagram name one pair of

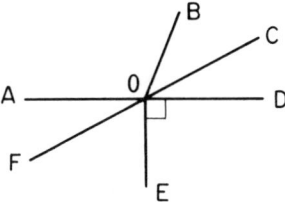

Fig. AT 8.7

(a) vertically opposite angles

(b) supplementary angles

(c) complementary angles

10. The figure shows four triangles with the sides marked a, b, and c. State in which triangles the equation

$$c^2 = b^2 + a^2$$

is valid (there may be more than one correct answer).

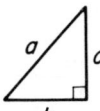 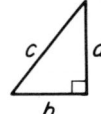 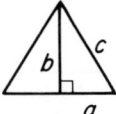 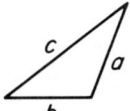

Fig. AT 8.8

11. For the triangle shown in the diagram which of the following is correct (there may be more than one correct answer)?

181

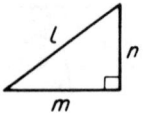

Fig. AT 8.9

(a) $l = m + n$
(b) $l^2 = m^2 + n^2$
(c) $m^2 = l^2 + n^2$
(d) $n^2 = l^2 + m^2$
(e) $m^2 = l^2 - n^2$

12. List I contains pairs of congruent triangles. List II gives the conditions of congruency. Match List II to List I by filling in the appropriate number in the boxes.

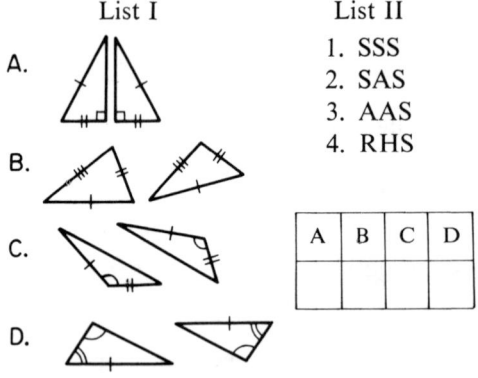

List I

A.

B.

C.

D.

List II

1. SSS
2. SAS
3. AAS
4. RHS

A	B	C	D

Fig. AT 8.10

13. Complete the following statements:

Two triangles are similar if their .... are equal.

The sides of two similar triangles are ....

The sides of two congruent triangles are ....

14. In the figure is shown two similar triangles. Which of the following is correct?

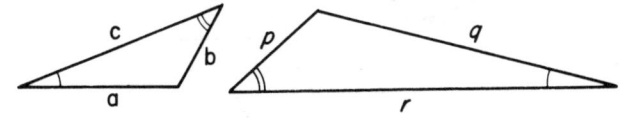

Fig. AT 8.11

(a) $\dfrac{a}{r} = \dfrac{b}{r}$

(b) $\dfrac{a}{r} = \dfrac{r}{b}$

(c) $\dfrac{b}{c} = \dfrac{r}{q}$

(d) $\dfrac{a}{q} = \dfrac{c}{r}$

15. Sketch the triangles with the following data:
    (a) an isosceles triangle ABC with the unequal angle A = 40°
    (b) a triangle ABC with $a = 60$ mm, $\angle$ C = 90°, $b = 100$ mm
    (c) a scalene triangle with angles of 40° and 60°, and the longest side of 50 mm

16. In a $\triangle$ ABC if $c^2 = a^2 + b^2$, then the triangle must be
    (a) equilateral
    (b) scalene
    (c) isosceles
    (d) right angled
    Select the correct answer.

17. Which of the following conditions do **not** prove congruency?
    (a) S.S.S.
    (b) A.A.A.
    (c) S.A.S.
    (d) A.A.S.
    (e) R.H.S.
    (f) A.S.S.

183

18. Determine the length p in each of the following:

(a)

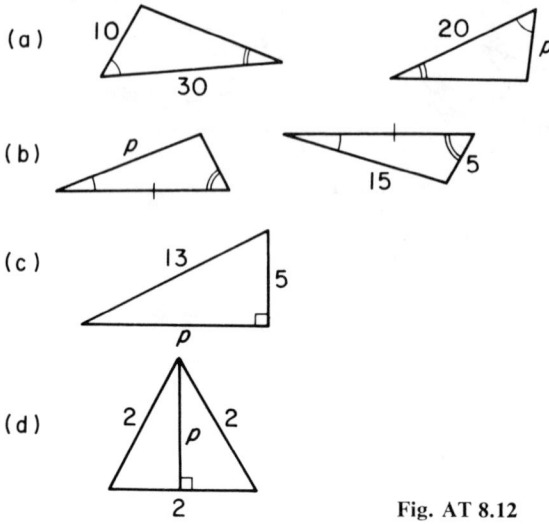

(b)

(c)

(d)

Fig. AT 8.12

19. Find the unknown angles marked $\alpha, \beta, \theta, \phi$ in the figures below.

(a)

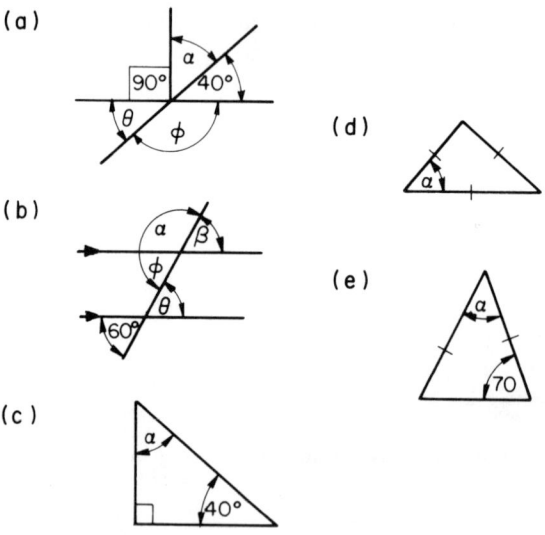

(b)

(c)

(d)

(e)

Fig. AT 8.13

20. Which of the following triangles is marked with the correct lengths of sides?

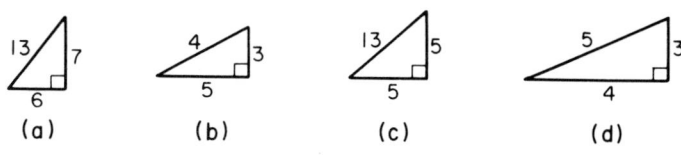

Fig. AT 8.14

21. The following pairs of triangles shaded ▨ and ▨ are congruent. State whether this **true** or **false** in each case.

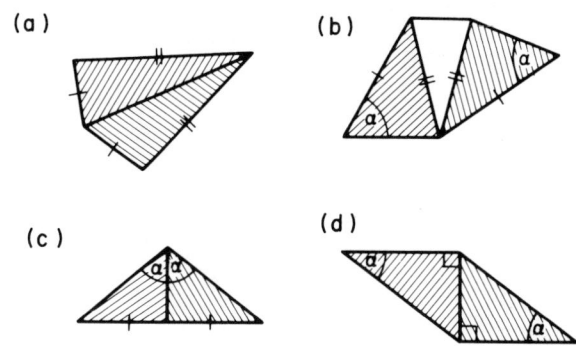

(a)

(b)

(c)

(d)

Fig. AT 8.15

22. In List I four right-angled triangles are drawn, with unknown sides marked x in each case. In List II the values of x are given. Match the correct value of x to the triangles by filling in the appropriate number in the vacant boxes.

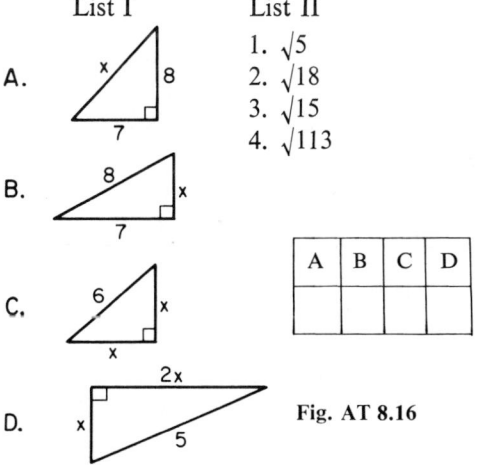

List I

A.

B.

C.

D.

List II

1. $\sqrt{5}$
2. $\sqrt{18}$
3. $\sqrt{15}$
4. $\sqrt{113}$

A	B	C	D

Fig. AT 8.16

185

23. Name all the parallel lines in the figure.

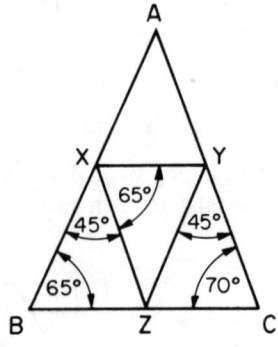

Fig. 8.17

9. Geometry 2

Objectives

After working through this chapter you should be able to

1. Identify the following parts of a circle: radius, diameter, circumference, semicircle, quadrant, sector, chord, segment, arc.
2. Calculate the value of the circumference or radius (or diameter) given one of the quantities.
3. State the relationship between the angles subtended at the centre and at the circumference of a circle, by any chord, including the special case of the diameter.
4. Determine unknown angles using the relationship in 3.
5. Identify the tangent to a circle.
6. State the value of the angle between a tangent and a radius at the point of contact with the circle.
7. State the relationship between two tangents drawn from a given point to a circle.
8. Solve problems involving the properties of tangents listed in 6 and 7.
9. State the value of the angle between a chord and a radius which bisects the chord.
10. Solve problems involving the properties of a chord listed in 9.
11. Explain the term radian.
12. Solve problems using the relationship between length of arc, radius and angle in radians.
13. Express angular rotation as a multiple of π radians.
14. Convert degrees into radians and vice versa using the relationship π rad $= 180°$.
15. Convert degrees and minutes to radians and vice versa using four-figure conversion tables.
16. State the properties of sides, angles and diagonals of the quadrilateral, parallelogram, rhombus, rectangle, square, trapezium, and regular hexagon.
17. Solve problems using these properties.

9.1 The circle

A circle is a plane (i.e., flat) figure bounded by a line which is always at the same distance from a fixed point.

Parts of a circle

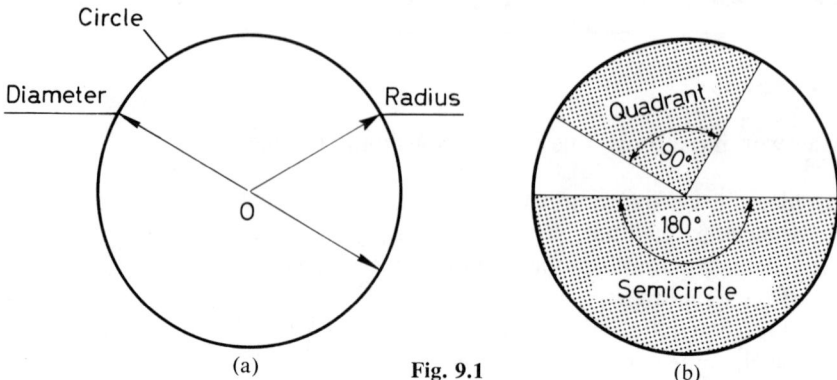

(a)   **Fig. 9.1**   (b)

Centre: This is the 'fixed point' in the definition. It is labelled point O in Fig. 9.1(a).

Radius (r): This is the distance from the fixed point to the curve.

Diameter (d): This is the width of the circle measured through the centre. It is obvious that radius = $\frac{1}{2}$ diameter.

Circumference (c): This is the distance around the circle, that is, its perimeter.

Semicircle: This is one-half of a circle.

Quadrant: This is one-quarter of a circle.

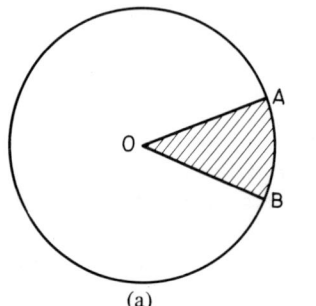
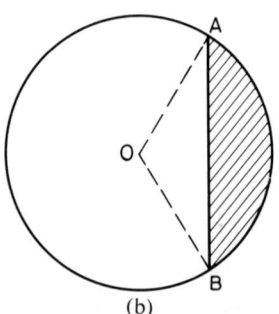

Fig. 9.2

(a)   (b)

A *sector* is the portion of a circle between two radii. The shaded portion in Fig. 9.2(a) is a minor sector and the unshaded portion a major sector.

A *chord* is a straight line joining two points on the circumference of a circle. In Fig. 9.2(b) the line AB is a chord.

Segments are the parts into which a circle is divided by a chord. In Fig. 9.2(b) the shaded part is a minor segment and the unshaded part the major segment.

An *arc* is a portion of the circumference of a circle. In Fig. 9.2(a) the short portion AB is called a minor arc, the longer portion AB being a major arc.

188

Circumference of a circle. It is found that if the circumference of any circle is measured and is divided by its diameter the answer is always equal to 3·142. This answer is correct only to three places of decimals, it is not an exact value. This ratio of circumference/diameter is called π (π is a Greek letter, pronounced 'pie'). Hence

$$\frac{\text{circumference}}{\text{diameter}} = \pi$$

or

$$\text{circumference} = \pi d$$

π is often taken as $\frac{22}{7}$ since $\frac{22}{7} = 3\cdot143$.

EXAMPLE 9.1 Find the circumference of a circle of diameter 70 mm.

$$\text{Circumference} = \pi d$$
$$= \pi \times 70$$
$$= \tfrac{22}{7} \times 70$$
$$= 220 \text{ mm}$$

EXERCISE 9.1

1. Find the circumference of the following circles,
   (a) 200 mm dia   (b) 0·46 m rad   (c) 1·7 m dia   (d) 1·62 m rad
2. A semicircular plate has a diameter of 2·4 m. Calculate the distance around the edge of the plate.
3. A base plate is in the form of a quadrant of a circle of radius 0·4 m. Find the distance around the edge of the plate.
4. A wheel of diameter 100 mm, rolling without slipping along a track, turns through 30 complete revolutions. Find the distance travelled along the track.
5. (a) The circumference of a bicycle wheel is 2·24 m. Calculate the number of complete turns it makes in travelling 1 km.

   (b) The minute hand of a clock is 56 mm long. Calculate how far the tip will travel between 4.34 a.m. and 5.16 a.m. (taking $\pi = \frac{22}{7}$).

   (NCTEC)
6. The circumference of a motor-car tyre is 1540 mm. Calculate the radius in m (taking $\pi = 3\frac{1}{7}$).
7. Find the maximum radius of a wheel which must rotate at least 10 times to cover a distance of 120 cm. Give your answer to the nearest mm.
8. A gear of diameter 50 mm has 40 teeth. Find the distance between adjacent teeth measured around the circumference.

9.2 Angles subtended by an arc of a circle

Theorem. *The angle subtended by an arc (or chord) of a circle at the centre is twice the angle subtended by the same arc (or chord) at the circumference,* i.e., in Fig. 9.3 $\angle$ AOC = 2$\angle$ ABC.

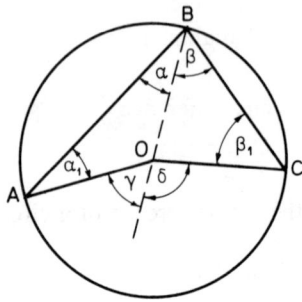

Fig. 9.3

Triangle AOB is isosceles since OB = OA (radii of the circle), therefore $\alpha = \alpha_1$
Now

$$\angle \text{ AOB} = 180 - 2\alpha \quad \text{(sum of angles in } \triangle \text{ AOB)}$$

and

$$\angle \text{ AOB} = 180 - \gamma \quad \text{(supplementary angles)}$$

Therefore

$$\angle \text{ AOB} = 180 - \gamma = 180 - 2\alpha$$

from which

$$\gamma = 2\alpha$$

By the same reasoning, in triangle OBC we obtain

$$\delta = 2\beta$$

Adding these two equations gives

$$\gamma + \delta = 2\alpha + 2\beta$$
$$= 2(\alpha + \beta)$$

that is,

$$\angle \text{ AOC} = 2\angle \text{ ABC}$$

A special case of this theorem occurs when AOC is a straight line, that is, when it is a diameter (Fig. 9.4).
Since $\delta = 180°$, from the theorem it follows that $\alpha = 90°$. This result may be stated as

The angle subtended at the circumference by a diameter is a right angle.

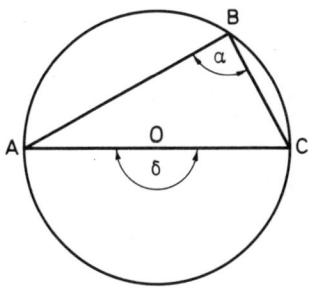

Fig. 9.4

EXAMPLE 9.2   In each of the following figures find the angle marked x. The centre of the circle is marked O in each case.

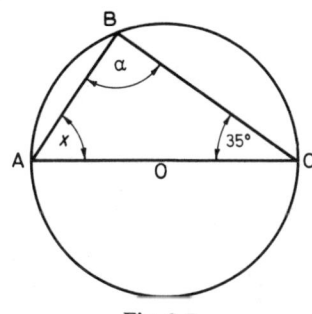

Fig. 9.5

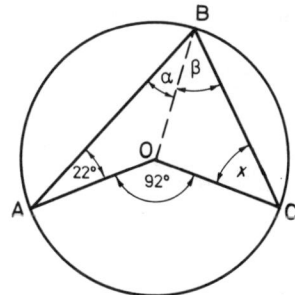

Fig. 9.6

(a) In Fig. 9.5 $\alpha = 90°$ (angle subtended by diameter).
    In $\triangle$ ABC

$$x+90+35 = 180 \qquad \text{(angle sum in triangle)}$$
$$x+125 = 180$$
$$x = 180-125$$
$$\therefore \quad x = 55°$$

(b) In Fig. 9.6 it is necessary to draw a construction line BO. Mark the angles α and β as shown. Now triangle AOB is isosceles because AO and OB are radii of the circle. Therefore

$$\alpha = 22°$$
$$92 = 2(\alpha+\beta) \quad \text{(angle at circumference)}$$
$$= 2\alpha+2\beta$$
$$= 44+2\beta$$

Therefore

$$2\beta = 92 - 44$$

$$= 48$$

$$\beta = 24$$

But $x = \beta$ since triangle OBC is isosceles (OB and OC are radii). Therefore

$$x = 24°$$

EXERCISE 9.2

1. In Figs. 9.7 and 9.8 determine the angles marked α. In both cases O is the centre of the circle.

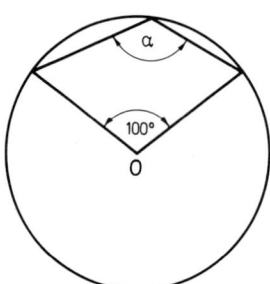

Fig. 9.7

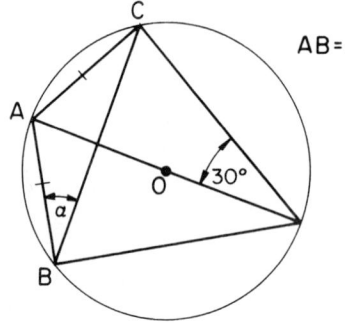

Fig. 9.8

2. Three holes A, B, and C are drilled in a plate on the circumference of a circle, centre O. If AB = AC and $\angle$ BOC is 76°, find the angles of the triangle ABC.

3. Three transistors A, B, and C are to be placed on a baseboard on the circumference of a circle in such a way that BC is a diameter of the circle. If AB = 0·06 m and BC = 0·10 m calculate the length of wire required to join up the three transistors.

4. BC is a diameter of a circle, centre O, and A is a point on the circumference such that $\angle$ ABC = 35°. Determine $\angle$ OAC.

5. Three transmitting stations A, B, and C are placed at equal distances from a television mast O. If $\angle$ AOC = 120°, $\angle$ BOC = 160°, determine the angles of the triangle ABC.

6. A steel plate has three holes A, B, and C drilled in it on the circumference of a circle, radius O. Find the angle ABC if $\angle$ AOB = 100° and $\angle$ BOC = 150°.

192

9.3 Properties of tangents

(a) A tangent is a straight line which touches a circle at one point only. In Fig. 9.9 XY is a tangent since it touches the circle at the point A only.

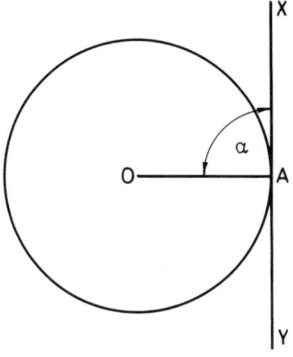

Fig. 9.9

(b) The angle between a tangent and a radius at the point of contact with the circle is 90°. In Fig. 9.9, α is the angle between the tangent and the radius, and this angle α is 90°.

(c) Two tangents, TA and TB (Fig. 9.10), may be drawn from a common point T to a circle. They have equal lengths, i.e. TA = TB. This may be proved by showing that triangles AOT and BOT are congruent.

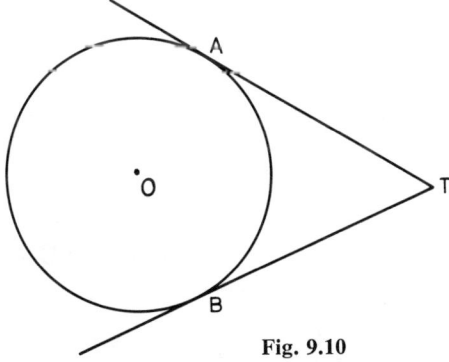

Fig. 9.10

EXAMPLE 9.3   A wheel of diameter 200 mm rests against a vertical surface as shown in Fig. 9.11. It is held in position shown by means of a rod OB, with the point A 500 mm from B. What length of rod is needed?

Now

$\angle$ OAB = 90°   (angle between tangent and radius)

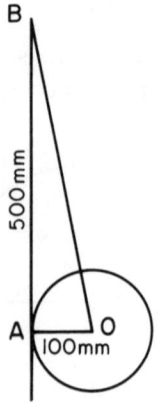

Fig. 9.11

Therefore

$$\triangle\ OAB\ \text{is right-angled}$$

From Pythagoras' theorem

$$OB^2 = OA^2 + AB^2$$
$$= 100^2 + 500^2$$
$$= 10\ 000 + 250\ 000$$
$$= 260\ 000$$

Therefore

$$OB = \sqrt{260\ 000}$$
$$= 510\ \text{mm}$$

9.4 Property of a chord

If a line is drawn from the centre of a circle perpendicular to a chord then the chord is bisected (Fig. 9.12), that is,

$$AC = CB$$

This is obviously true because $\triangle$ AOB is isosceles (AO = OB = radius).

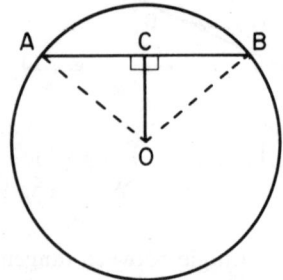

Fig. 9.12

194

EXAMPLE 9.4   A steel cam is to be circular with a flat edge AB as shown in Fig. 9.13.

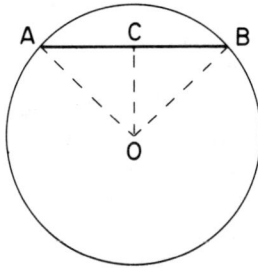

Fig. 9.13

The flat edge is to be 120 mm long and must be at a distance of 80 mm from the centre of the circle. What is the required radius of the cam?

In $\triangle$ AOC

$$AC = \tfrac{1}{2} \times 120 = 60 \text{ mm} \text{ (property of a chord)}$$

and

$$OC = 80 \text{ mm}$$

By Pythagoras' theorem, since $\angle$ C = 90,

$$OA^2 = OC^2 + AC^2$$

$$= 80^2 + 60^2$$

$$- 10\ 000$$

Therefore

$$OA = \sqrt{10\ 000} = 100 \text{ mm}$$

that is, the required radius OA is 100 mm.

EXERCISE 9.3

1. A chord subtends an angle of 90° at the centre of a circle of radius 0·65 m. Calculate the length of the chord.
2. A flat of width 180 mm is to be ground on a shaft of diameter 260 mm. Find the depth of material removed.
3. A steel shaft of diameter 180 mm is supported by an endless loop of rope, which passes around a hook H, in such a way that the centre of the shaft is 150 mm vertically below H. Calculate the length of the portion of rope from H to the point of contact with the shaft.

4. A tangent TA touches a circle of radius 50 mm at the point A. From A a chord AB is drawn at an angle of 60° to TA. Calculate the length of the chord.
5. A cylindrical steel roller of diameter 120 mm is to rest in a horizontal slot such that the depth of the roller within the slot is 30 mm. Calculate the minimum width of the slot.
6. A 10·0 m long ladder leans against a building which is in the form of a hemispherical dome of diameter 8·0 m. If the foot of the ladder rests on the ground at a point 3·0 m from the base of the building, calculate the length of the ladder protruding beyond its point of contact with the building.
7. The mouth of a road tunnel is in the form of a major segment of diameter 8·2 m. If the width of the horizontal base is 6·0 m, what is the maximum headroom of the tunnel?
8. One pulley drives another pulley of equal diameter by means of a belt passing over the two pulleys. If the diameter of each pulley is 80 mm and the pulley centres are 100 mm apart, find the length of the belt.

9.5 Radian measure and length of arc

An angle may be measured in **degrees** or **radians**. Expressing angles in radians simplifies many calculations; in others the calculations cannot be evaluated unless the angles are reduced to radians.

A **radian** is defined as the angle subtended at the centre of a circle by an arc equal in length to the radius. In Fig. 9.14 the length of the arc AB is r. The angle subtended at O by AB is thus 1 radian (1 rad).

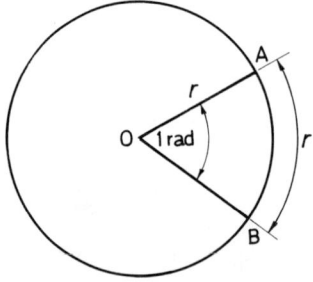

Fig. 9.14

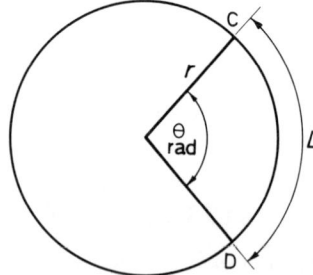

Fig. 9.15 ·

Length of arc of a circle. In Fig. 9.15 the arc CD subtends and angle θ rad at the centre O of the circle of radius r. From the definition for a radian we have that 1 rad is subtended by an arc of length r and θ rad is subtended by an arc of length $r\theta$. Therefore, if the length of arc is L, we have

$$L = r\theta$$

or

arc = radius × angle in radians

The relationship between degrees and radians. If we think of an arc as the complete circumference of a circle, its length is $2\pi r$. Using the equation $L = r\theta$ we can calculate the angle subtended at the centre. It is

$$\theta = \frac{L}{r} = \frac{2\pi r}{r} = 2\pi$$

However, the angle at the centre is one complete revolution, or 360°. Therefore

$$2\pi \text{ rad} = 360°$$

or

$$\pi \text{ rad} = 180°$$

From this relationship

$$1 \text{ rad} = \frac{180}{\pi} = 57° \ 18'$$

Conversion between degrees and radians may be carried out using this relationship, or using conversion tables.

Conversion between degrees and radians using the relationship

$$\pi \text{ rad} = 180°$$

is quite convenient if the angular rotation in radians is expressed as a multiple of π radians. For example, since $360° = 2\pi$ rad then $720° = 4\pi$ rad. Conversely 6π must equal $3 \times 360° = 1080°$.

For more difficult conversions involving minutes and decimals it is better to use conversion tables.

EXAMPLE 9.5

(a) Without using tables convert (i) $\pi/6$ rad to degrees, and (ii) 80° to radians.
(b) Using tables convert (i) 185° 27' to radians and (ii) 0·2249 radians to degrees.

(a) (i) From the above relationship

$$\pi \text{ rad} = 180°$$

$$\frac{\pi}{6} \text{ rad} = \frac{180}{6} = 30°$$

(ii)

$$180° = \pi \text{ rad}$$

$$80° = \frac{\pi}{180} \times 80 \text{ rad}$$

$$= \tfrac{4}{9}\pi \quad \text{or} \quad 1·395 \text{ rad}$$

197

(b) (i) We cannot read 185° 27′ directly from tables. It is split up into

$$180° + 5° 27′$$

We know from the preceding discussion that

$$180° = \pi \text{ rad} = 3.1416 \text{ rad}$$

From tables, the relevant section of which is reproduced in Fig. 9.16, it is seen that

$$5° 24′ = 0.0942 \text{ rad}$$

Add   3′   9   (Mean Differences)

$$5° 27′ = 0.0951 \text{ rad}$$

Hence

$$185° 27′ = 3.1416 + 0.0951$$

$$= 3.2367 \text{ rad}$$

DEGREES TO RADIANS

Degrees	0' 0°.0	6' 0°.1	12' 0°.2	18' 0°.3	24' 0°.4	30' 0°.5	36' 0°.6	42' 0°.7	48' 0°.8	54' 0°.9	Mean Differences 1 2 3	4 5
0	0000	0017	0035	0052	0070	0087	0105	0122	0140	0157	3 6 9	12 15
1	0175	0192	0209	0227	0244	0262	0279	0297	0314	0332	3 6 9	12 15
2	0349	0367	0384	0401	0419	0436	0454	0471	0489	0506	3 6 9	12 15
3	0524	0541	0559	0576	0593	0611	0628	0646	0663	0681	3 6 9	12 15
4	0698	0716	0733	0750	0768	0785	0803	0820	0838	0855	3 6 9	12 15
5	0873	0890	0909	0935	0942	0960	0977	0995	1012	1030	3 6 9	12 15
6	1047	1065	1082	1100	1117	1134	1152	1169	1187	1204	3 6 9	12 15
7	1222	1239	1257	1274	1292	1309	1326	1344	1361	1379	3 6 9	12 15
8	1396	1414	1431	1449	1466	1484	1501	1518	1536	1553	3 6 9	12 15
9	1571	1588	1606	1623	1641	1658	1676	1693	1710	1728	3 6 9	12 15
10	1745	1763	1780	1798	1815	1833	1850	1868	1885	1902	3 6 9	12 15
11	1920	1937	1955	1972	1990	2007	2025	2042	2060	2077	3 6 9	12 15
12	2094	2112	2129	2147	2164	2182	2199	2217	2234	2251	3 6 9	12 15
13	2269	2286	2304	2321	2339	2356	2374	2391	2409	2426	3 6 9	12 15
14	2443	2461	2478	2496	2513	2531	2548	2566	2583	2601	3 6 9	12 15

Fig. 9.16

(ii) In order to convert 0.2249 rad into degrees the nearest smaller number to it is obtained on the table. This number is 0.2234 which is boxed in the table. This number is seen to be equivalent to 12° 48′. The remainder of the number 0.2249 is 0.0015 rad. From the mean differences column this is seen to be equivalent to 5′:

$$0.2234 \text{ rad} = 12° 48′$$

Add   15   5′   (Mean Differences)

$$\therefore \quad 0.2249 \text{ rad} = 12° 53′$$

EXAMPLE 9.6 Figure 9.17 shows a pulley P driven by a belt. The belt leaves the pulley at points X and Y. The angle of contact of the belt (known as the angle of lap) is 120°. If the radius of the pulley is 100 mm calculate the length of the belt always in contact with the pulley.

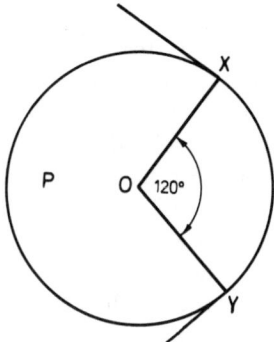

Fig. 9.17

$$180° = \pi \text{ rad}$$

$$\therefore \quad 120° = \frac{\pi}{180} \times 120 \text{ rad}$$

$$= \tfrac{2}{3}\pi \text{ rad}$$

Now, arc length, $L = r\theta$

$$= 100 \times \tfrac{2}{3}\pi$$

$$= 209 \text{ mm}$$

EXERCISE 9.4

1. Without using tables, change the following angles from radians to degrees:

   (a) π (b) 2π (c) $\dfrac{\pi}{2}$ (d) $\dfrac{3\pi}{2}$ (e) $\dfrac{\pi}{3}$ (f) $\dfrac{\pi}{4}$

   (g) $\dfrac{2\pi}{3}$ (h) $\dfrac{3\pi}{4}$ (i) $\dfrac{5\pi}{6}$ (j) $\dfrac{5\pi}{3}$ (k) 3π (l) $\dfrac{7\pi}{3}$

2. Without using tables change the following angles from degrees into radians, leaving the answers in terms of π:
   (a) 30° (b) 9° (c) 10° (d) 225° (e) 330° (f) 420°

3. Use the conversion factor to convert the following angles from radians to degrees and minutes:
   (a) 0·5 (b) 1·2 (c) 3·0 (d) 0·17 (e) 5·6

4. Using the conversion factor convert the following angles into radians:
   (a) 12° (b) 18° (c) 14° (d) 106° (e) 326°
5. Use tables to convert the following angles into radians:
   (a) 27° (b) 42° 15′ (c) 84° 2′ (d) 185° 50′ (e) 302° 25′
6. Use tables to convert the following angles from radians into degrees and minutes:
   (a) 0·4 (b) 0·1172 (c) 1·213 (d) 1·0082
7. Two radii of length 30 mm are inclined to one another at an angle of 42°. Determine the lengths of the minor and major arcs.
8. Two spokes of a wheel are inclined at 72° to each other. If they are 1·12 m apart at the rim when measured around the rim, find the radius of the wheel.
9. An arc of a circle of radius 25 mm is 59 mm long. What is the angle subtended at the centre in degrees?
10. Rope is wound around a pulley of diameter 300 mm. If the pulley turns through an angle of 260° what length of rope unwinds from the pulley?
11. A wheel of diameter 500 mm rolls without slipping along a horizontal surface. If the wheel rotates 2·4 revs how far does the wheel move horizontally?
12. A belt passes around a pulley of diameter 100 mm. If 160 mm of the belt is in contact with the pulley find the angle at the centre of the pulley subtended by this portion of the belt. Give your answer in degrees and minutes.
13. The shaft of a dynamo rotates through an angle of 6π radians in 1 s. How many revolutions will this shaft make in 1 min.

9.6 Quadrilaterals

A quadrilateral is a plane figure bounded by four straight lines (Fig. 9.18).

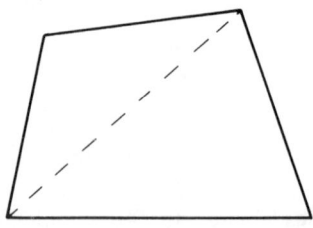

Fig. 9.18

It is seen that a diagonal (that is a line joining opposite corners) divides a quadrilateral into two triangles.

There are five quadrilaterals of special interest.

200

(a) Parallelogram

A parallelogram is a quadrilateral with both pairs of opposite sides parallel (Fig. 9.19).

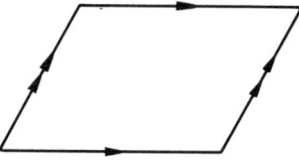

Fig. 9.19

By drawing and using congruent triangles it can be shown that the parallelogram has the following properties:
(a) The opposite sides are equal in length.
(b) The opposite angles are equal.
(c) The diagonals bisect one another.
(d) Each diagonal divides the parallelogram into two congruent triangles.

(b) Rhombus

A rhombus is a parallelogram with all its sides equal in length (Fig. 9.20).

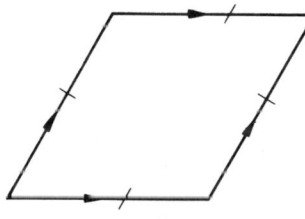

Fig. 9.20

A rhombus has all the properties of a parallelogram plus
(a) the diagonals bisect one another at 90°,
(b) the diagonals bisect the corner angles of the rhombus.

(c) Rectangle

A rectangle is a parallelogram with all its angles equal to 90° (Fig. 9.21).

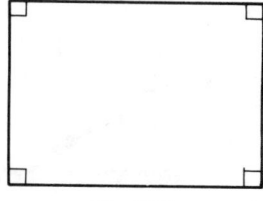

Fig. 9.21

A rectangle has all the properties of a parallelogram plus the fact that the diagonals bisect one another.

(d) Square

A square is a rectangle with all its sides equal in length (Fig. 9.22).

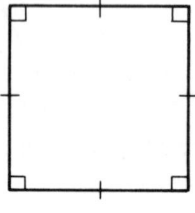

Fig. 9.22

A square has all the properties of the rhombus and the rectangle.

(e) Trapezium

A trapezium is a quadrilateral with one pair of parallel sides (Fig. 9.23).

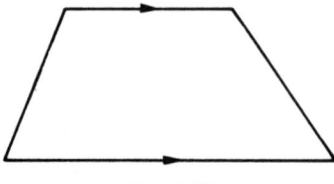

Fig. 9.23

9.7 Regular hexagon

A hexagon is a plane figure bounded by six straight lines. The regular hexagon shown in Fig. 9.24 has all its sides and angles equal.

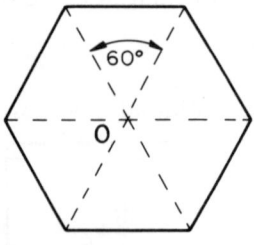

Fig. 9.24

All the triangles in the figure are equilateral. Therefore the corner points of the hexagon lie on a circle of centre O.

EXAMPLE 9.7   A nut in the form of a regular hexagon is to be machined from a round bar of diameter 100 mm. What size of spanner will be required to fit the 'across flats' dimension X of this nut, to the nearest mm?

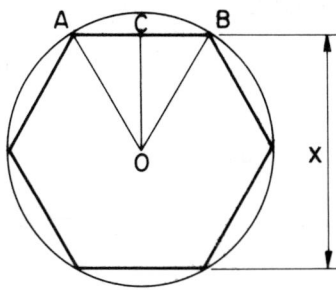

Fig. 9.25

In Fig. 9.25

$$AB = OB = 50 \text{ mm} \quad (\triangle \text{ AOB is equilateral})$$

and

$$BC = \tfrac{1}{2}AB \qquad \text{(property of chord)}$$

Therefore

$$BC - 25$$

By Pythagoras' theorem

$$OC^2 = OB^2 - BC^2$$

$$= 2500 - 625$$

$$= 1875$$

Hence

$$OC = 43\cdot3 \text{ mm}$$

$$\text{'Across flats' dimension} = 2 \times OC$$

$$= 2 \times 43\cdot3$$

$$= 87 \text{ mm} \quad \text{(to the nearest mm)}$$

EXERCISE 9.5

1. (a) By drawing a diagonal to a quadrilateral prove that the angles of a quadrilateral add up to 360°.
   (b) Hence calculate the unknown angles α, β in Fig. 9.26.

203

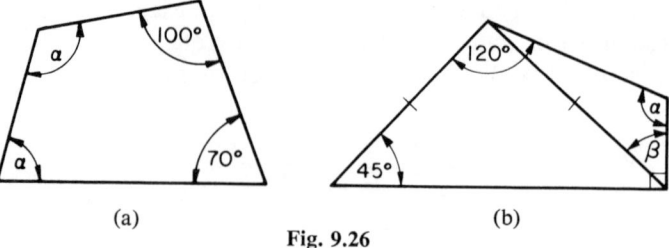

(a)   (b)

Fig. 9.26

2. Evaluate α and β in the two diagrams in Fig. 9.27.

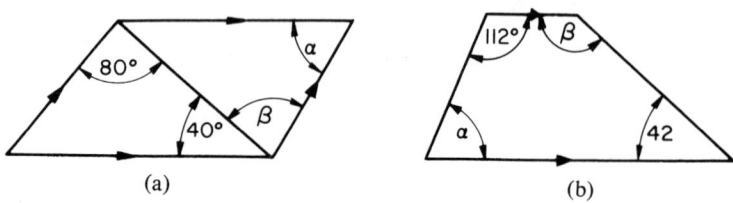

(a)   (b)

Fig. 9.27

3. An aluminium sheet is to be cut in the form of a rhombus. The distances across the corners of the rhombus are 600 mm and 800 mm. Find the lengths of the sides of the sheet.

4. Using congruent triangles prove that
   (a) the diagonals of a parallelogram bisect one another;
   (b) the diagonals of a rhombus bisect one another at right angles,
   (c) the diagonals of a rhombus bisect the corner angles.

Assessment test 9

1. The shaded area in the diagram is
   (a) a quadrant
   (b) a sector
   (c) a segment
   (d) an arc

   Select **two** correct answers.

Fig. AT 9.1

2. The ratio of the two shaded areas is
   (a) 2 : 1
   (b) 4 : 1
   (c) 3 : 1
   (d) 4 : 3

   Select the correct answer.

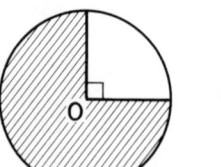

Fig. AT 9.2

204

3. The length of the arc AB in the diagram is π. Select the correct value of circumference and radius from the following list:

(a) 2π
(b) 3π
(c) 2
(d) 3
(e) 6
(f) 6π

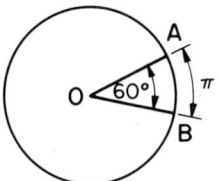

Fig. AT 9.3

4. Complete the following statements:

(a) The angle between a tangent and a radius is ....

(b) Tangents to a circle from a common point are of .... length.

(c) A radius drawn .... to a chord bisects the chord.

(d) The angle at the circumference of a circle is .... the angle at the centre subtending the same arc.

5. What is the value of the angle α. Select the correct answer from the list.

(a) 6 rad
(b) $\frac{3}{2}$ rad
(c) $\frac{2}{3}$ rad
(d) $\frac{2}{3}\pi$ rad

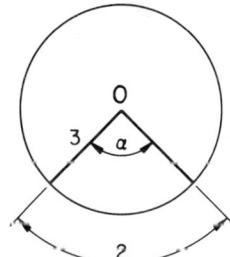

Fig. AT 9.4

6. A circle has a circumference of 44. Taking π as $\frac{22}{7}$ select the correct value of the diameter from

(a) 7
(b) 14
(c) $\frac{7}{2}$
(d) $\frac{1}{7}$

7. Complete the following statements:

(a) $\frac{1}{2}\pi$ radians is the same as .... degrees of rotation.

(b) Using tables, 0·49 rad = ...°.

(c) An arc of length 0·8 m subtends an angle of .... rad at the centre of a circle of diameter 1·0 m.

(d) Using tables, 124° 13′ = ... rad.

8. In the following diagrams identify the angles which are 45°.

(a)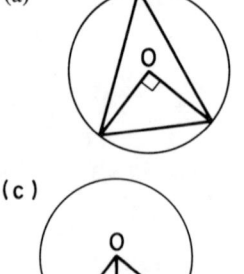

(b)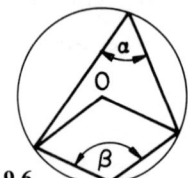

(c)

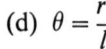

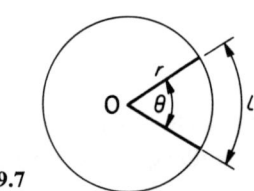

Fig. AT 9.5

9. In the diagram two angles are marked as α and β. Are these angles ·
   (a) equal?
   (b) supplementary?
   (c) corresponding?
   (d) complementary?

Fig. AT 9.6

10. What is the correct relationship between l, r, and θ shown in the diagram?
    (a) $\theta = r - 1$
    (b) $\theta = rl$
    (c) $\theta = \dfrac{l}{r}$
    (d) $\theta = \dfrac{r}{l}$

Fig. AT 9.7

11. The circumference of a circle is 8π. What is its radius?
    (a) 8π (b) 4 (c) 8 (d) 2π

12. Complete the following statements:
    (a) A regular hexagon has six .... sides.
    (b) The corner points of a regular hexagon lie on a ....
    (c) The angle between any two adjacent sides of a regular hexagon is ....°.
    (d) The triangles formed by drawing the diagonals between opposite points of a regular hexagon are ....

206

13. How much is

    (a) $\frac{1}{6}\pi$ rad in degrees?

    (b) 900° in radians?

    (c) $\frac{1}{4}\pi$ in degrees?

    (d) 120° in radians?

14. In the following diagrams mark in the angles that are 90°. State a reason why the angle is 90° in each case.

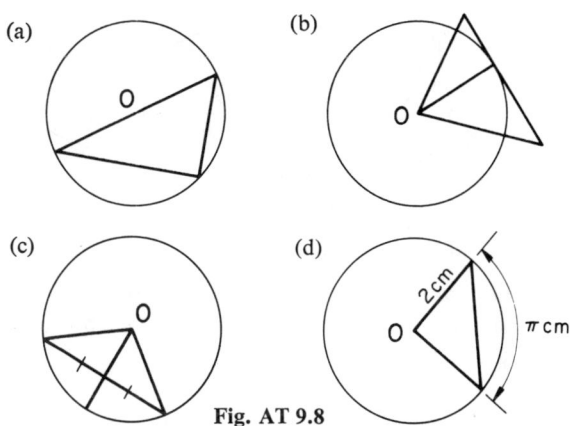

Fig. AT 9.8

15. The diagram shows a circle, centre O. List I contains the labelled parts of the diagram. List II contains the names of these labelled parts. Match the correct name to each part by filling in the appropriate numbers in the empty boxes

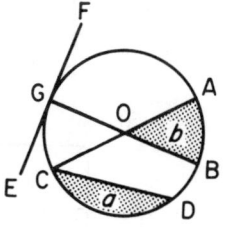

Fig. AT 9.9

List I	List II
A. EF	1. arc
B. CD	2. sector
C. OA	3. radius
D. area a	4. chord
E. BG	5. diameter
F. AB	6. tangent
G. area b	7. segment

A	B	C	D	E	F	G

16. In each of the circles shown the centre is labelled O. Find the angle α in each case.

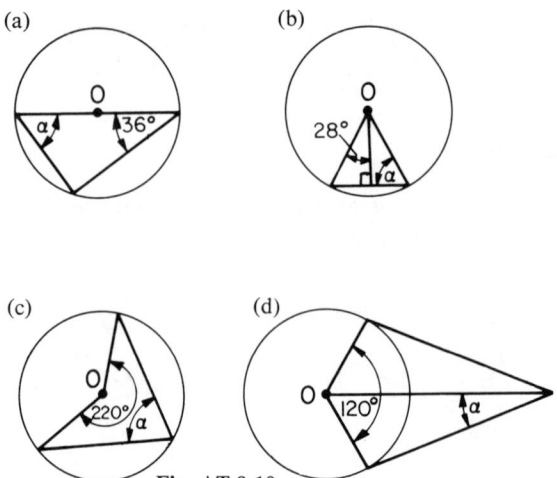

Fig. AT 9.10

17. Complete the following statements:

    (a) A quadrilateral has .... straight sides.

    (b) If the diagonals of a parallelogram bisect each other at 90°, then it must be a ....

    (c) The opposite sides of a parallelogram are .... and ....

    (d) A rhombus with the corner angles equal to 90° is a ....

18. Name each of the following figures:

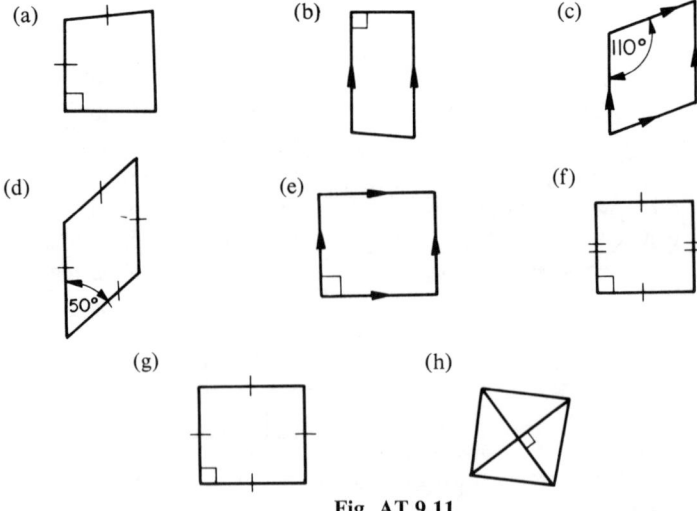

Fig. AT 9.11

10. Area and volume

Objectives

After working through this chapter you should be able to

1. Use given formulae to calculate the areas of squares, rectangles, parallelograms, triangles, trapeziums, and simple plane figures made up from these parts.
2. For known areas use the formulae in 1 to find the dimensions of an unknown side or height of a square, rectangle, parallelogram, or triangle.
3. Determine the perimeter of the figures listed in 1.
4. Use the relationship between area, radius (or diameter) of a circle to calculate whichever of these quantities is not known.
5. Determine the area of an annulus and a semicircle.
6. Calculate the areas of the figures which are made up from parts listed in 1 and circles, semicircles, and quadrants.
7. Use given formulae to calculate volumes and surface areas of rectilinear prisms, including the cylinder.
8. Use given formulae to calculate volumes of spheres, cones, square- and hexagonal-based pyramids.
9. Calculate the volumes of composite figures made up from the simple volumes listed in 8 and 9.
10. Calculate the volume of hollow figures in the form of a cylinder or sphere.
11. Calculate the heights or diameter of a cylinder or cone given its volume and one of these quantities.
12. Calculate the side of the base of a square-based prism, given its volume and height.

10.1 Perimeter and area

(a) Perimeter

The perimeter of a plane figure is the distance around the outside boundary of the figure.

(b) Area

Area is the measure of the size of a surface. It is concerned with surface coverage only and not with thickness. The area of a square of sides 1 unit long can be taken as a unit of area, that is, area $= 1 \times 1 = 1$ (Fig. 10.1).

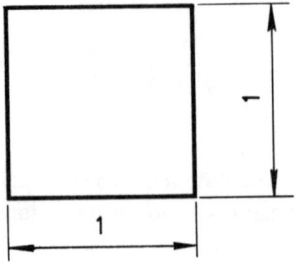

Fig. 10.1

To find the area of any plane (i.e., flat) surface it is therefore necessary to find how many such squares of unit size there are in the area.

To determine the area of any figure it is necessary to multiply length by length. Hence

$$\text{unit of area} = \text{length} \times \text{length}$$

$$= \text{squared length or length}^2$$

Units of area are square metres, written as m^2, square millimetres mm^2, etc. Since $1\ m = 10^3\ mm$, then

$$1\ m^2 = 10^3 \times 10^3\ mm^2 = 10^6\ mm^2, \quad \text{and} \quad 1\ km^2 = 10^6\ m^2.$$

10.2 Areas of plane rectilinear figures

A rectlinear figure is a figure bounded by straight lines, for example, a triangle, a quadrilateral, etc.

(a) Square (Fig. 10.2)

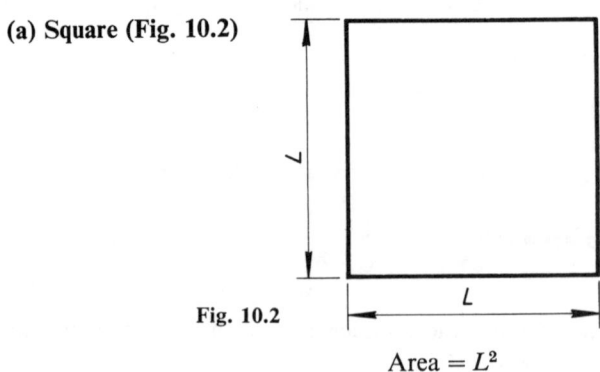

Fig. 10.2

$$\text{Area} = L^2$$

(b) Rectangle (Fig. 10.3)

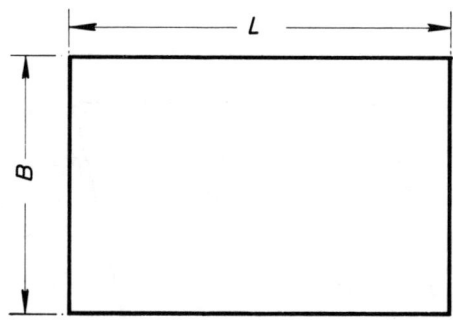

Fig. 10.3

Area $= L \times B$

(c) Parallelogram

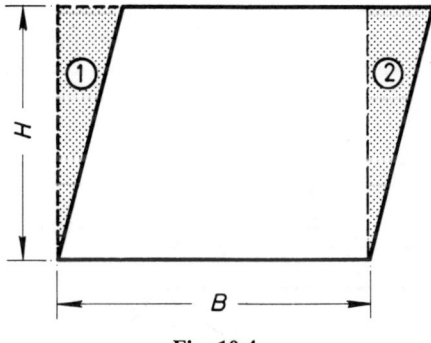

Fig. 10.4

Area of rectangle = unshaded area + shaded area ①.

Area of parallelogram = unshaded area + shaded area ②.

Since the shaded areas ① and ② are equal, then

Area of parallelogram = area of rectangle

$= $ base $\times$ height

Area $= BH$

(d) Triangle

A diagonal divides a parallelogram into two congruent triangles (see Example 8.7).

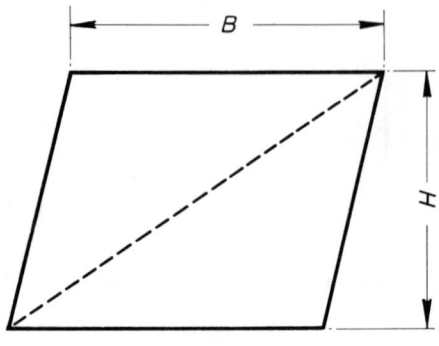

Fig. 10.5

Therefore,

$$\text{Area of triangle} = \tfrac{1}{2} \times \text{area of parallelogram}$$
$$= \tfrac{1}{2} \times \text{base} \times \text{height}$$
$$\text{Area} = \tfrac{1}{2}BH$$

Note: Any side may be considered as the base, the height is then the perpendicular distance from this chosen base to the opposite vertex or corner.

(e) Trapezium

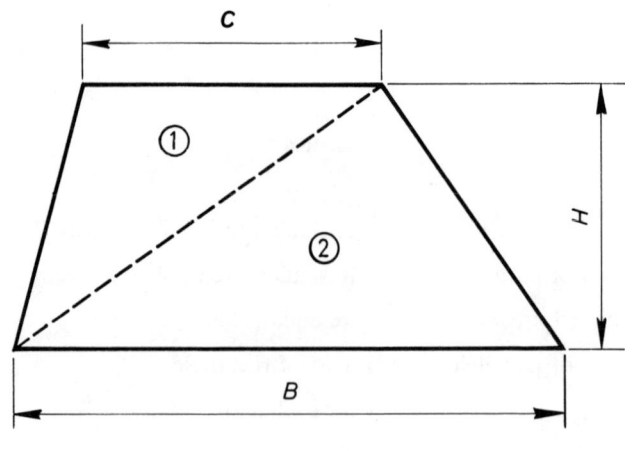

Fig. 10.6

$$\text{Area of trapezium} = \text{area of triangle} \enspace \textcircled{1} + \text{area of triangle} \enspace \textcircled{2}$$

$$= \tfrac{1}{2}CH + \tfrac{1}{2}BH$$

that is, $$\text{Area} = \tfrac{1}{2}(C+B)H$$

$$\text{area of trapezium} = \tfrac{1}{2} \text{ (sum of parallel sides)} \times \text{(perpendicular}$$
$$\text{distance between the parallel sides)}$$

EXAMPLE 10.1   Determine the area of a trapezium with parallel sides of lengths 40 mm and 70 mm if the vertical distance between these parallels is 30 mm.

$$\text{Area} = \tfrac{1}{2}(\text{sum of parallels}) \times (\text{distance between them})$$

$$= \tfrac{1}{2}(40+70) \times 30$$

$$= 55 \times 30$$

$$= 1650 \text{ mm}^2$$

EXAMPLE 10.2   Determine the area of sheet metal required to make an open rectangular box 280 mm wide, 0·4 m long, and 60 mm high, as shown in Fig. 10.7.

Note: All units of length must be the same, i.e., change 0·4 m to 400 mm.

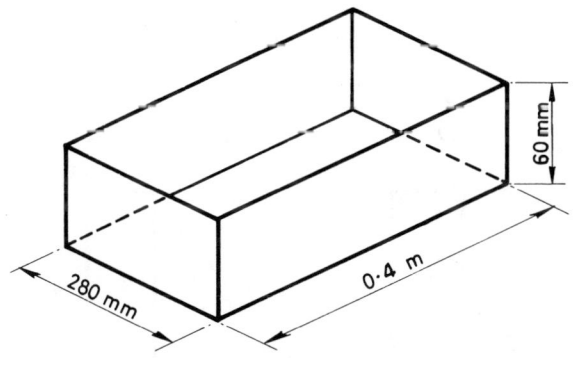

Fig. 10.7

$$\text{Area of two sides} = 2 \times 60 \times 400 = \quad 48\ 000 \text{ mm}^2$$

$$\text{Area of two ends} = 2 \times 60 \times 280 = \quad 33\ 600 \text{ mm}^2$$

$$\text{Area of base} \quad\quad = 280 \times 400 \quad = \underline{112\ 000 \text{ mm}^2}$$

$$193\ 600 \text{ mm}^2$$

$$\text{Sheet metal required} = 194 \times 10^3 \text{ mm}^2.$$

EXERCISE 10.1

1. Calculate the area and perimeter of
   (a) a square with sides 15 mm long;
   (b) a rectangle of length 250 mm and width 40 mm.
2. Determine the area and perimeter of the figures shown. (Figs. 10.8, 10.9, 10.10.)

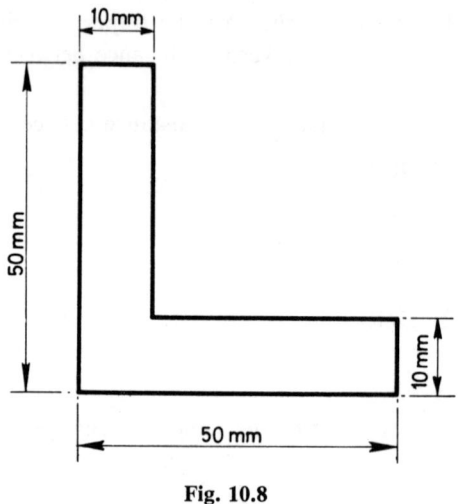

Fig. 10.8

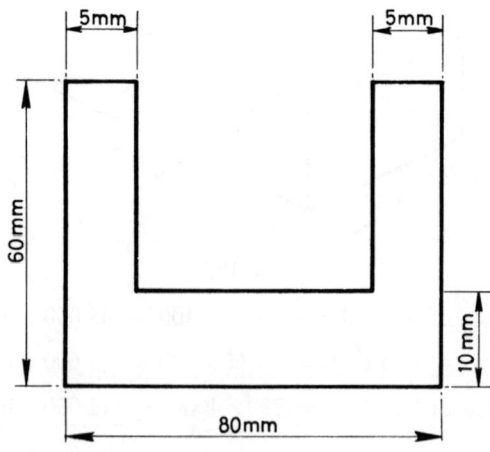

Fig. 10.9

214

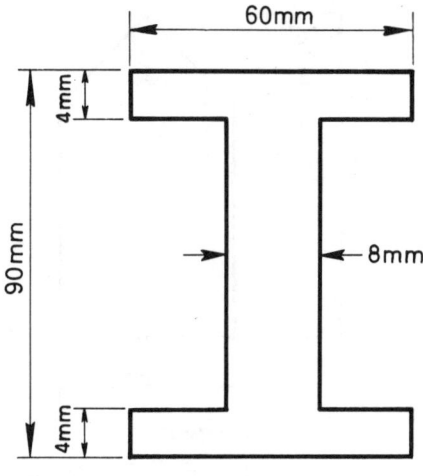

Fig. 10.10

3. Find the areas of the triangles with the following dimensions:
   (a) base 0·40 m, vertical height 0·82 m;
   (b) base 318 mm, vertical height 0·200 m;
   (c) an isosceles triangle with base 80 mm and equal sides of 50 mm;
   (d) an equilateral triangle with sides of length 100 mm.
4. Calculate the area of the steel plates with the dimensions shown. (Figs. 10.11 and 10.12.)

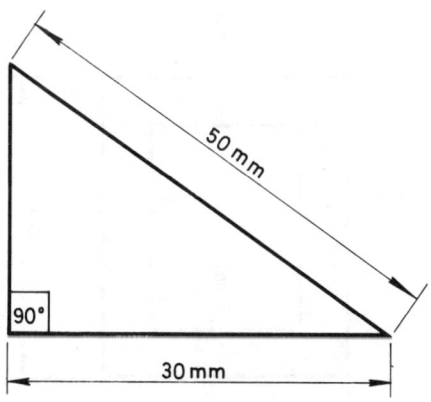

Fig. 10.11

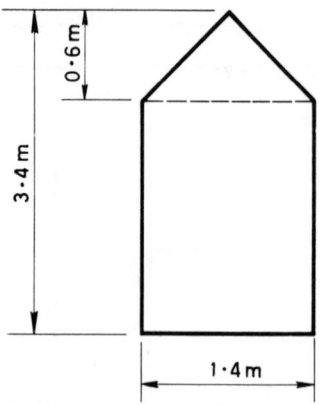

Fig. 10.12

5. (a) The longest side of a parallelogram is 150 mm. If the longest sides are 80 mm apart, find the area of the parallelogram.
   (b) A parallelogram has an area of 0·064 m². If one side of the parallelogram is 140 mm long, calculate the vertical height of the parallelogram measured from this side.

6. (a) Calculate the side of a square having an area of 4800 mm².
   (b) If the sides of a rectangular slot are in the ratio 3 : 1 and its area is 54 mm², determine the dimensions of the slot.

7. The cross-sectional area of a V-belt is 50 mm². If its width is 15 mm what is the depth of the section?

8. A steel tube 8·0 m long has a cross-section as shown in Fig. 10.13. Calculate the inside and outside surface area of the tube, neglecting the area of the end faces.

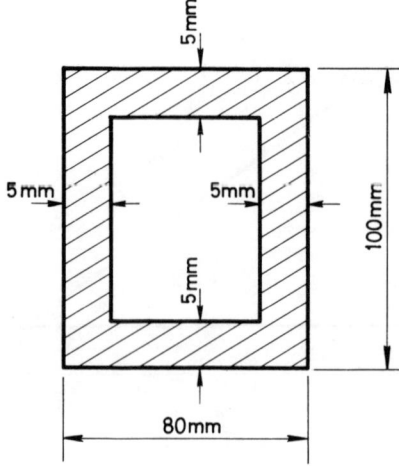

Fig. 10.13

9. Two V-grooves, machined in a flat surface, have the same width, and depths of 8 mm and 12 mm. Compare the cross-sectional areas of the grooves.

10.3 Areas of circular figures

(a) Area of a circle

$$\text{Area of circle} = \pi r^2, \quad \text{where } r = \text{radius.}$$

This formula can be written in terms of the diameter, d, where $\frac{1}{2}d = r$:

$$\text{Area} = \pi\left(\frac{d}{2}\right)^2 = \pi\frac{d^2}{4}$$

EXAMPLE 10.3 Find the area of a circle of diameter 70 mm.

$$\text{Area of circle} = \frac{\pi d^2}{4} = \frac{22}{\cancel{7}} \times \frac{1}{4} \times \cancel{70}^{10} \times 70$$

$$= 3850 \text{ mm}^2$$

EXAMPLE 10.4 The end face of a boiler is made from a flat circular plate. If the area of the plate is 27 m² calculate the boiler diameter.

$$\text{Area of plate} = \frac{\pi d^2}{4}$$

$$\therefore \quad 27 = \frac{\pi d^2}{4}$$

Multiply by 4:

$$27 \times 4 = \pi d^2$$

Divide by π:

$$\frac{27 \times 4}{\pi} = d^2$$

$$\therefore \quad d^2 = \frac{27 \times \cancel{4}^{2} \times 7}{\cancel{22}_{11}} = 34\cdot36$$

$$\therefore \quad d = \sqrt{34\cdot36}$$

$$\therefore \quad \text{Boiler diameter} = 5\cdot86 \text{ m}$$

(b) Annulus

An annulus is the space between two concentric circles, i.e., two circles with the same centre, see Fig. 10.14.

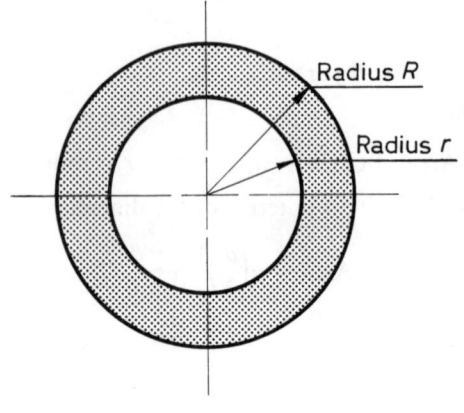

Fig. 10.14

An example of an annulus is the face of an ordinary washer. The area of an annulus is simply the difference in area between the two circles,

$$\text{Area of annulus} = \pi R^2 - \pi r^2$$
$$= \pi(R^2 - r^2)$$

EXAMPLE 10.5   A 60 mm diameter hole is punched centrally through a thin metal sheet, of diameter 200 mm. What is the area of the annulus so formed?

$$\text{Area} = \pi(R^2 - r^2)$$
$$= \tfrac{22}{7}(100^2 - 30^2)$$
$$= \tfrac{22}{7} \times 9100$$
$$= 28\ 600\ \text{mm}^2 \quad \text{or} \quad 0.0286\ \text{m}^2$$

EXERCISE 10.2

1. Calculate the area of the following circles:
   (a) 200 mm dia   (b) 0·46 m rad
   (c) 1·7 m dia   (d) 1·62 m rad
2. A semicircular plate has a diameter of 2·14 m. Calculate its area.
3. A base plate is in the form of a quadrant of a circle of radius 0·4 m. Find the area of the plate.

4. A circular plate has an area of 1300 mm². Determine its diameter and hence calculate the circumference.
5. Calculate the shaded areas and external perimeters in Figs. 10.15 to 10.21.

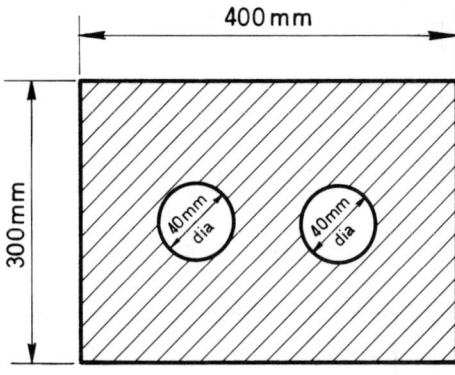

Fig. 10.15

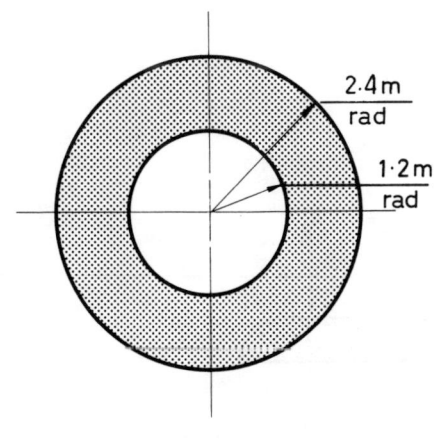

Fig. 10.16

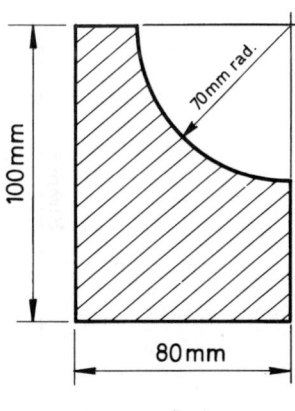

Fig. 10.17

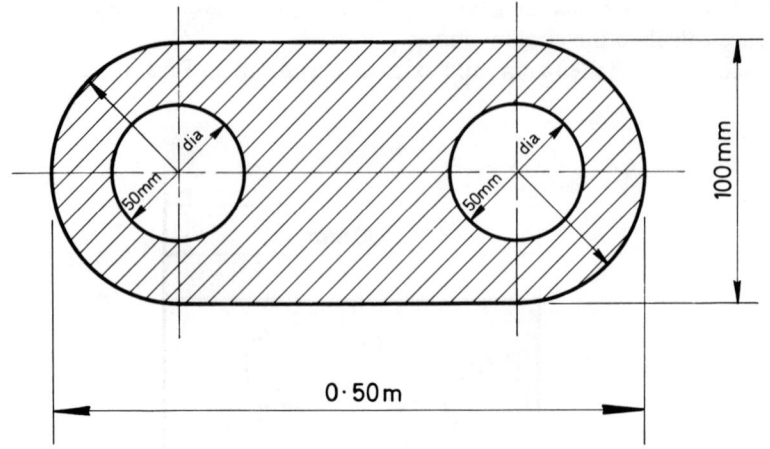

Fig. 10.18

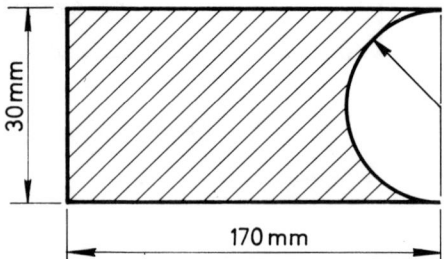

Fig. 10.19

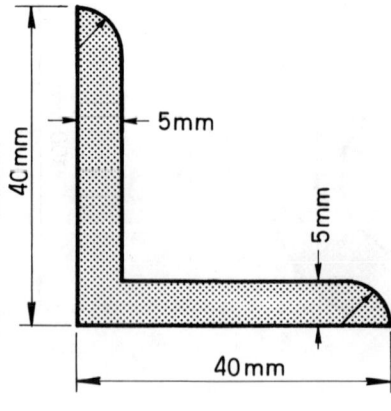

Fig. 10.20

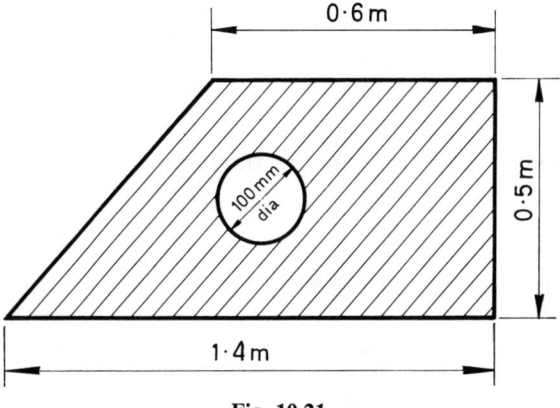

Fig. 10.21

6. A path of uniform width 1 m is to be placed around a circular pond of diameter 50 m. Determine the area of the path.
7. The end plate of a steel boiler is a flat circular plate of diameter 2·4 m. Holes are drilled in it as follows: two 450 mm dia fire-tube holes and twenty water-tube holes 30 mm dia. What is the final area of the plate?
8. The four corners of a rectangular cover, 500 mm by 300 mm, are rounded off to a radius of 25 mm. Determine the final area of the cover.

10.4 Volumes of prisms

The volume of any solid is a measure of the space occupied by the material from which the solid is made. The volume of a cube of side 1 unit long can be taken as a unit of volume, i.e., volume $= 1 \times 1 \times 1 = 1$ (see Fig. 10.22).

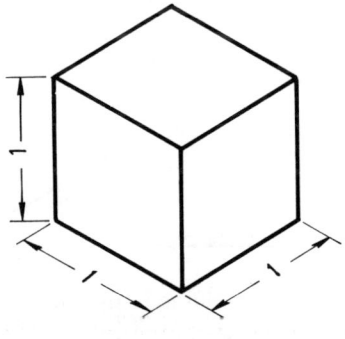

Fig. 10.22

To find the volume of any solid it is therefore necessary to find how many such cubes of unit size are in the solid.

221

Units of volume

Since volume = length × length × length, then volume = cubic length or length^3, i.e., units of volume are cubic millimetres (mm^3), cubic metres (m^3), etc., where 1 m^3 = 10^9 mm^3.

10.5 Volumes and surface areas of prisms

Definition: A prism is a solid with a constant cross-section.

(a) Cube

Consider a cube of side 3 units (Fig. 10.23). It is seen that this cube can be

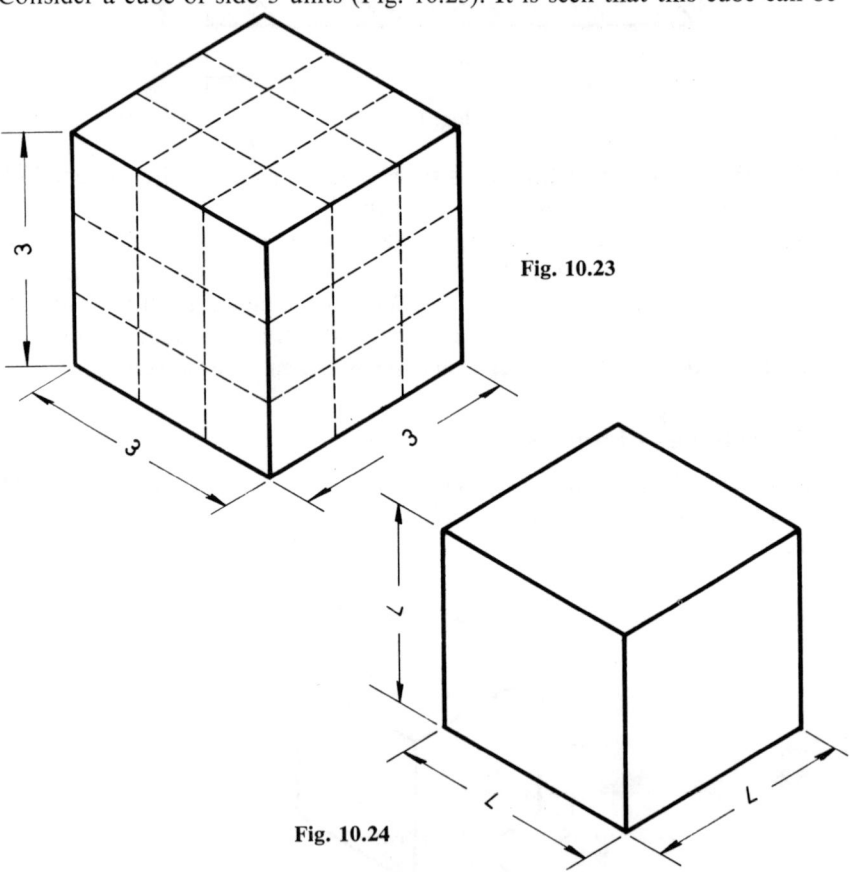

Fig. 10.23

Fig. 10.24

split up into 27 unit cubes. Its volume is therefore 27. This answer can be obtained by multiplying three sides of the cube together,

i.e., volume of cube = 3 × 3 × 3 = 27.

For any cube of side L units (Fig. 10.24), volume of cube = $L × L × L$, or

$$\text{Volume} = L^3$$

(b) Rectangular prism

This has a cross-section in the shape of a rectangle.
Consider a rectangular prism with the dimensions shown in Fig. 10.25. By

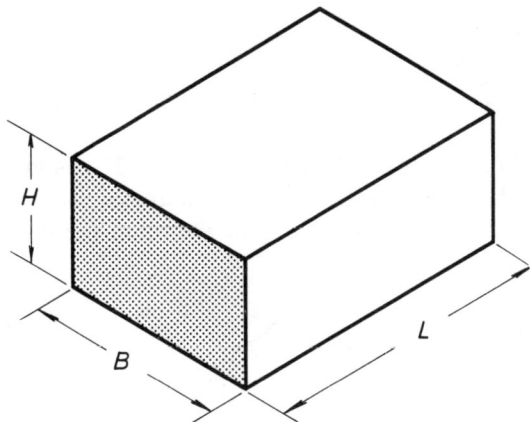

Fig. 10.25

dividing the prism into unit cubes the volume can be shown to be:

$$\text{Volume} = \text{breadth} \times \text{height} \times \text{length}$$

$$= BHL$$

but $\qquad BH = \text{area of end face,}$

therefore $\qquad \text{Volume} = \text{area of end face} \times \text{length}$

(c) Any prism

This formula is true for any prism irrespective of the shape of the base (Fig. 10.26), so

$$\text{Volume of any prism} = (\text{area of end face}) \times (\text{length of prism})$$

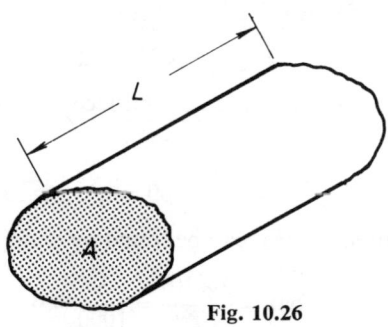

Fig. 10.26

223

Surface area of a prism

The area of the sides of a prism is known as the lateral surface area and can be found from the formula:

Lateral surface area = (perimeter of base) × (length of prism)

To find the total surface area of the prism, the areas of the end faces must also be added to the lateral area.

EXAMPLE 10.6. A steel bar, shown in Fig. 10.27, 2·4 m long, is in the form of a triangular prism. Its section is a right-angled isosceles triangle with its equal sides of length 50 mm. Determine the volume of the bar.

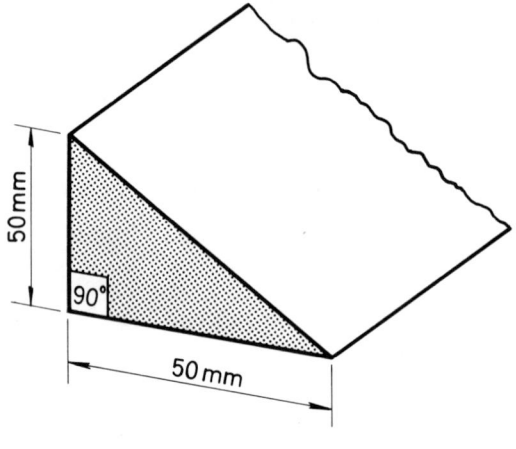

Fig. 10.27

Since the end face is triangular, then

$$\text{Area of end face} = \tfrac{1}{2}\ \text{base} \times \text{height}$$

$$= \tfrac{1}{2} \times 50 \times 50 = 1250\ \text{mm}^2$$

$$\therefore \quad \text{Volume of bar} = \text{area of end face} \times \text{length}$$

$$= 1250 \times 2400$$

$$= 3\ 000\ 000\ \text{mm}^3 \quad \text{or} \quad 3 \times 10^6\ \text{mm}^3$$

If this answer is required in m³ it can be found by dividing by 10^9, i.e.,

$$\text{Volume of bar} = \frac{3 \times 10^6}{10^9} = \frac{3}{1000} = 0\text{·}003\ \text{m}^3$$

224

EXERCISE 10.3

1. Find the volume of a cube of side 40 mm.
2. Find the volume of the following rectangular blocks:
   (a) length 70 mm, height 30 mm, width 15 mm;
   (b) length 1·3 m, height 0·50 m, width 40 mm.
   Put your answers in standard form.
3. A storage bin 2·0 m long with vertical sides has a section in the form of a trapezium as shown in Fig. 10.28. What is the capacity of the bin?

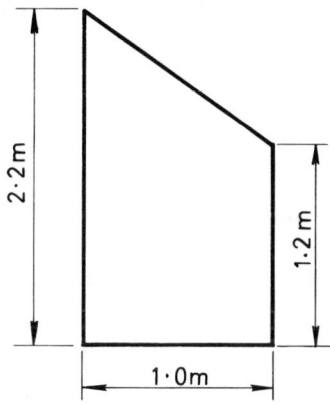

Fig. 10.28

4. A prism of length 2·0 m has a volume of 3·5 m³. Calculate its cross-sectional area.
5. What is the volume of an iron casting with a uniform triangular cross section and the following dimensions: length of casting 2·2 m, base of triangle 300 mm, height of triangle 80 mm?
6. Calculate the volume of a mild steel bar, 4 m long, if it has an I-section, in which the top flange is 20 mm wide, the bottom flange is 30 mm wide, the depth of the web is 25 mm, and the thickness of the web and the flanges is 4 mm.
7. Determine the volume of the V-block which has a thickness of 50 mm and a section as shown in Fig. 10.29.
8. What is the volume per metre run of copper bar with a uniform cross-section in the form of a regular hexagon with sides 12 mm long?

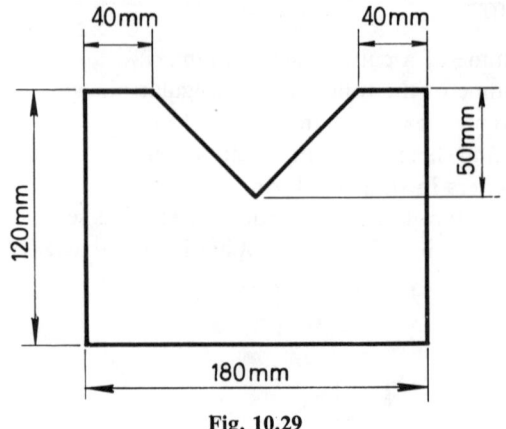

40mm   40mm

50mm

120mm

180mm

Fig. 10.29

10.6 Volume and surface area of a cylinder

A cylinder is a prism with a circular base. The term 'right' cylinder is sometimes used which means that the axis of the cylinder is perpendicular to the base. Unless otherwise stated it is always assumed that any cylinder is a right cylinder. Consider a cylinder with a base radius r and height (i.e., the distance measured along its axis) h (Fig. 10.30).

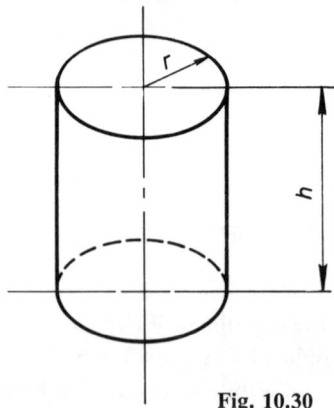

r

h

Fig. 10.30

Surface area

The surface area around the side of the cylinder is known as the curved surface area. If a sheet of paper was cut until it fitted exactly (without overlapping) around the curved surface of the cylinder it would be found that the paper would be rectangular in shape as shown in Fig. 10.31.

226

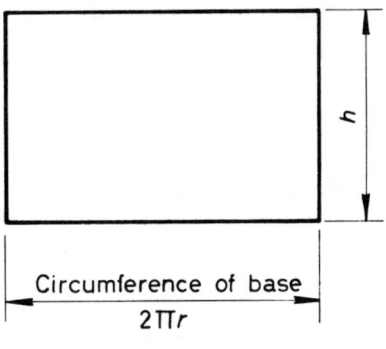

Circumference of base

$2\pi r$

Fig. 10.31

Hence the curved surface area of the cylinder $=$ area of rectangle

$$= \text{perimeter of base} \times \text{height}$$

$$= 2\pi rh \quad \text{or} \quad \pi dh$$

Note: If the area of the ends has to be considered, as for example with a solid cylinder or a hollow cylinder with a base and a lid, then these areas, which are circular, must be added to the curved surface area to give the total surface area.

Volume

Since a cylinder is a prism then

$$\text{Volume of cylinder} = \text{area of base} \times \text{height}$$

$$= \pi r^2 h \quad \text{or} \quad \frac{\pi d^2 h}{4}$$

EXAMPLE 10.7 Find the volume and total surface area of a cylinder which has a diameter of 100 mm and a length of 60 mm.

$$\text{Volume of cylinder} = \frac{\pi}{4} d^2 h$$

$$= 3 \cdot 142 \times \frac{100^2}{4} \times 60$$

$$= 471\ 000 \text{ mm}^3$$

227

$$\text{Curved surface area} = \pi dh$$

$$= 3{\cdot}142 \times 100 \times 60$$

$$= 18\ 900 \text{ mm}^2$$

$$\text{Area of two circular ends} = 2\pi r^2$$

$$= 2 \times 3{\cdot}142 \times 2500$$

$$= 15\ 700 \text{ mm}^2$$

$$\therefore \quad \text{Total surface area} = 18\ 900 + 15\ 700$$

$$= 34\ 600 \text{ mm}^2$$

Note: **The volume of material in a pipe is volume of outer cylinder − volume of inner cylinder.**

EXERCISE 10.4

1. Find the volume and curved surface area of the following cylinders:
    (a) 120 mm dia 100 mm high
    (b) 0·27 m dia 1·54 m high
    (c) 248 mm rad 0·500 m high

2. A round bar of diameter 18 mm has flat ends and a length of 3·4 m. Calculate the total surface area of this bar.

3. A cylinder has a length of 100 mm and a volume of 8×10^6 mm³. Determine its diameter and hence calculate its curved surface area.

4. The curved surface area of a cylinder is 880 mm² and its diameter is 14 mm. Calculate its height (taking $\pi = \frac{22}{7}$).

5. Calculate the volume of the casting shown in Fig. 10.32 (taking $\pi = \frac{22}{7}$).

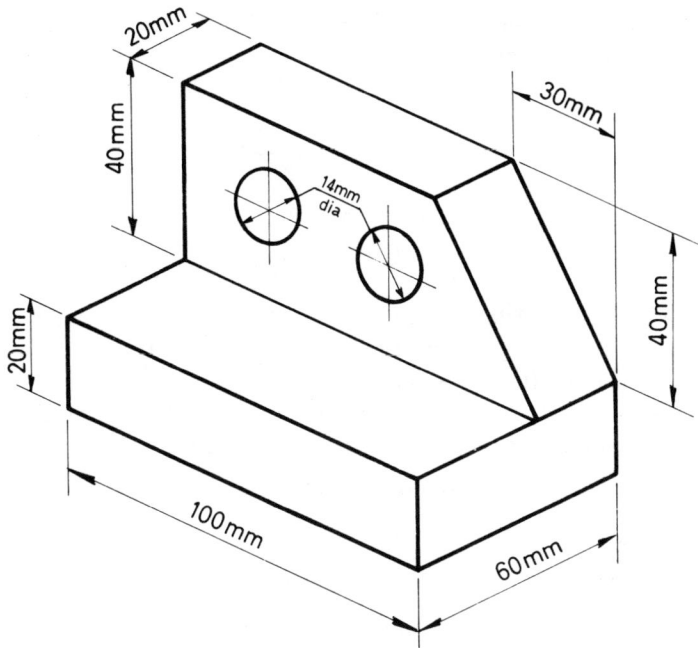

Fig. 10.32

6. Calculate the area of sheet metal required to make two open cylindrical-shaped ships' funnels of height 3·60 m and diameter 2·00 m allowing 5% extra material for overlapping.
7. A cylindrical brass tube has an outside diameter of 75 mm and an inside diameter of 50 mm. What is the volume of material on this tube per metre run?
8. A steel spacer ring 0·2 m thick has an outside diameter of 1·4 m and an inside diameter of 1·2 m. Find the volume of steel required to make this ring.
9. A solid steel shaft of diameter 125 mm and 1·2 m long is to be replaced by a hollow steel shaft of the same length and 250 mm outside diameter. Calculate the inside diameter of the hollow shaft, if the volume of material in both shafts is the same.

10.7 Volume of a sphere

A sphere is a solid figure which is circular in shape viewed from any direction, e.g., billiard ball.

If r = radius of the sphere then

$$\text{Volume of sphere} = \tfrac{4}{3}\pi r^3$$

229

EXAMPLE 10.8 Find the volume of a sphere of radius 20 mm. Put your answer in standard form.

$$\text{Volume} = \tfrac{4}{3}\pi r^3 = \tfrac{4}{3} \times 3 \cdot 142 \times 20^3$$

$$= \frac{3 \cdot 142 \times 4 \times 8000}{3}$$

$$= 33\ 500\ \text{mm}^3$$

$$= 3 \cdot 35 \times 10^4\ \text{mm}^3$$

EXAMPLE 10.9 Determine the volume of plastic in cubic metres required to make a hollow ball with an outside diameter of 210 mm and an inside diameter of 168 mm (taking $\pi = \tfrac{22}{7}$).

The volume of material required can be found by subtracting the volume of the spherical air space inside the ball from the outside volume of the ball

$$\text{Inside volume} = \tfrac{4}{3}\pi r^3 = \tfrac{4}{3} \times \tfrac{22}{7} \times 84^3$$

$$= 88 \times 84^2 \times 4$$

$$= 2\ 480\ 000\ \text{mm}^3$$

$$\text{Outside volume} = \tfrac{4}{3} \times \tfrac{22}{7} \times 105^3$$

$$= 88 \times 105^2 \times 5$$

$$= 4\ 850\ 000\ \text{mm}^3$$

$$\therefore\quad \text{Volume of plastic} = 4\ 850\ 000 - 2\ 480\ 000$$

$$= 2\ 370\ 000\ \text{mm}^3$$

$$= \frac{2\ 370\ 000}{10^9}\ \text{m}^3 = 0 \cdot 00237\ \text{m}^3$$

EXERCISE 10.5

1. Calculate the volume of a sphere of radius (a) 10 mm, (b) 0·04 m, and (c) 2·2 m.
2. A spherical pressure vessel has an internal volume of 8 m³. What is the internal diameter of the vessel?
3. A storage tank is in the form of a cylinder with a flat base and a hemi-spherical top. The tank has a diameter of 500 mm and the cylindrical part of the tank is 1·0 m long. Calculate the tank capacity.
4. Calculate the capacity of a cylindrical tank with hemispherical ends if the overall length of the tank is 3·0 m and the diameter is 1·2 m.
5. A steel rivet is in the form of a cylinder with a hemispherical head. If the overall length of the rivet is 80 mm, the head diameter is 40 mm, and the shank diameter is 20 mm, find the volume of the rivet.

6. A copper sphere of diameter 50 mm is melted down and drawn out into a wire 200 m long. Calculate the diameter of the wire.

10.8 Volume of a pyramid and a cone

Note: The centre of a rectangle, square, or regular hexagon is the point at which the diagonals cross each other.

(a) Pyramids

Definition: A pyramid is a solid figure with a rectilinear base and sloping sides meeting at a point (Figs. 10.33 and 10.34).

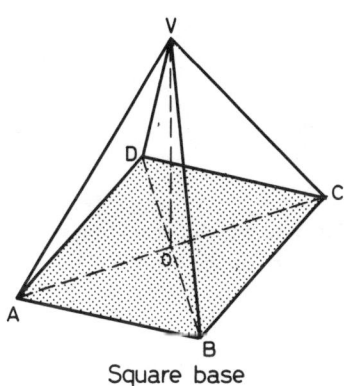

Square base

Fig. 10.33

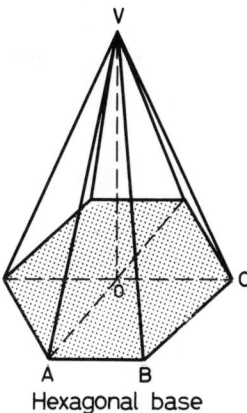

Hexagonal base

Fig. 10.34

The point at which the sides meet is called the vertex. If the vertex is vertically above the centre of the base the pyramid is called a right pyramid. In all cases:

O is the centre of the base
V is the vertex
VO is the 'height' of the pyramid, i.e., vertical height
Volume of pyramid = $\frac{1}{3}$ area of base × height

EXAMPLE 10.10 Find the volume of a pyramid with a square base 60 mm by 60 mm and 40 mm high, as shown in Fig. 10.35.

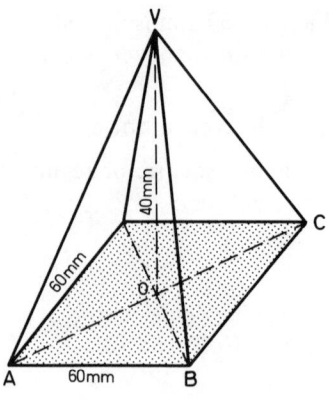

Fig. 10.35

$$\text{Volume of pyramid} = \tfrac{1}{3} \text{ area of base} \times \text{height}$$

$$= \tfrac{1}{3} \times 60 \times 60 \times 40$$

$$= 48\,000 \text{ mm}^3$$

(b) Cones

Definition: A cone is a solid figure with a circular base which tapers to a point. A right cone has its vertex vertically above the centre of the base.

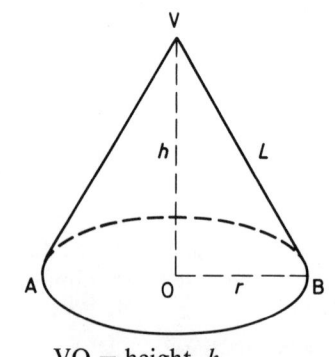

Fig. 10.36

In Fig. 10.36   VO = height, h

OB = base radius, r

VB = slant height, L

In $\triangle$ VOB, by Pythagoras, $r^2 + h^2 = L^2$

$$\text{Volume of cone} = \tfrac{1}{3} \text{ area of base} \times \text{height}$$

$$= \tfrac{1}{3}\pi r^2 h$$

Note: The volume of a cylinder is $\pi r^2 h$, i.e., a cone has a volume which is a third of a cylinder with the same radius and height.

EXAMPLE 10.11 Calculate the volume of a right cone with a base diameter of 140 mm and a slant height of 250 mm (take $\pi = \frac{22}{7}$).

Fig. 10.37

In the triangle in Fig. 10.37 by Pythagoras

$$h = \sqrt{(250^2 - 70^2)}$$
$$= \sqrt{5\,7600}$$
$$= 240 \text{ mm.}$$
$$\text{Volume} = \tfrac{1}{3}\pi r^2 h$$
$$= \tfrac{1}{3} \times \tfrac{22}{7} \times 70 \times 70 \times 240$$
$$= 1{\cdot}23 \times 10^6 \text{ mm}^3$$

EXERCISE 10.6

1. Calculate the volume of the following pyramids:
   (a) square base, of side 80 mm and height 30 mm;
   (b) regular hexagon base, of side 10 mm and height 40 mm.
2. Determine the volume of the cones with the following dimensions:
   (a) base diameter 30 mm and height 70 mm
   (b) base radius 5·0 mm and slant height 13 mm
   (c) height 140 mm and semi-vertical angle 30°
   (d) base diameter 12 mm and vertical angle 90°.
3. A brass plumb bob is in the form of a hemisphere with a cone on the top. Calculate the volume of the bob if the diameter of the hemisphere is 14 mm and the overall height is 37 mm.
4. Calculate the volume of metal required to make the lathe centre shown in Fig. 10.38.

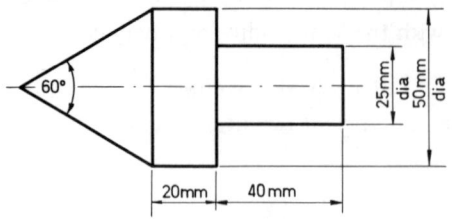

Fig. 10.38

5. In order to balance a high-speed steel disk a hole is drilled near its rim as shown in Fig. 10. 39. Find the volume of metal removed.

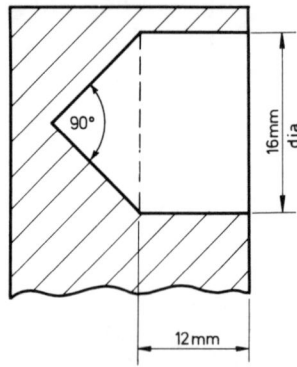

Fig. 10.39

6. A pyramid with a square base of side 100 mm has the same height as a cone of equal volume. What is the base radius of the cone?

Assessment test 10

1. In the diagram four plane figures are shown. Three of them have something in common. What is it, and which is the odd one out?

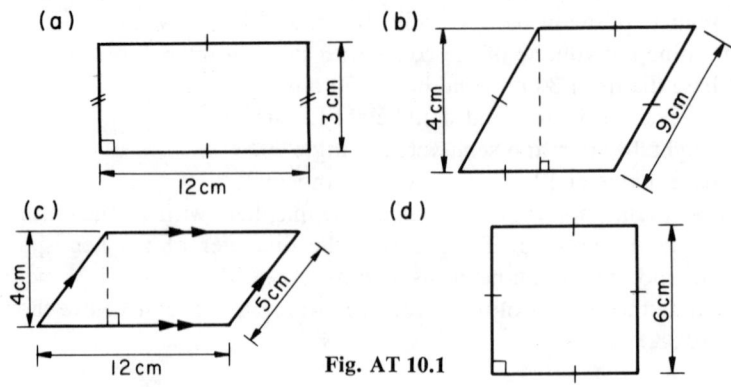

Fig. AT 10.1

2. The diagram consists of a circle, centre O, and a shaded square of area 9 cm². What is the area of the circle?

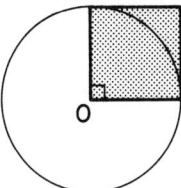

(a) 3π cm²
(b) 6π cm²
(c) 9π cm²
(d) $81\ \pi$ cm²

Fig. AT 10.2

3. The diagram shows a square (shaded) drawn on the diameter of a semi-circle, centre O. The area of the square is 16 cm². What is the area of the semicircle?

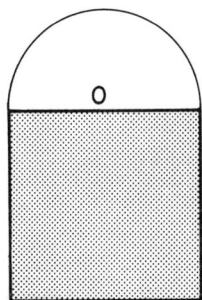

(a) 2π cm²
(b) 4π cm²
(c) 8π cm²
(d) 16π cm²

Fig. AT 10.3

4. What is the value of the shaded area shown in the diagram? Select the correct answer from the following:

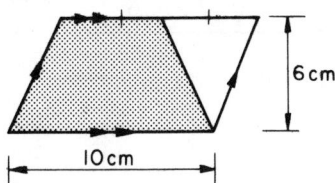

6 cm

10 cm

(a) 60 cm²
(b) 30 cm²
(c) 45 cm²
(d) 70 cm²

Fig. AT 10.4

5. List I shows four shaded areas. List II gives the values of these areas. Match the correct value to the figure, by filling in the appropriate numbers in the empty boxes.

List I

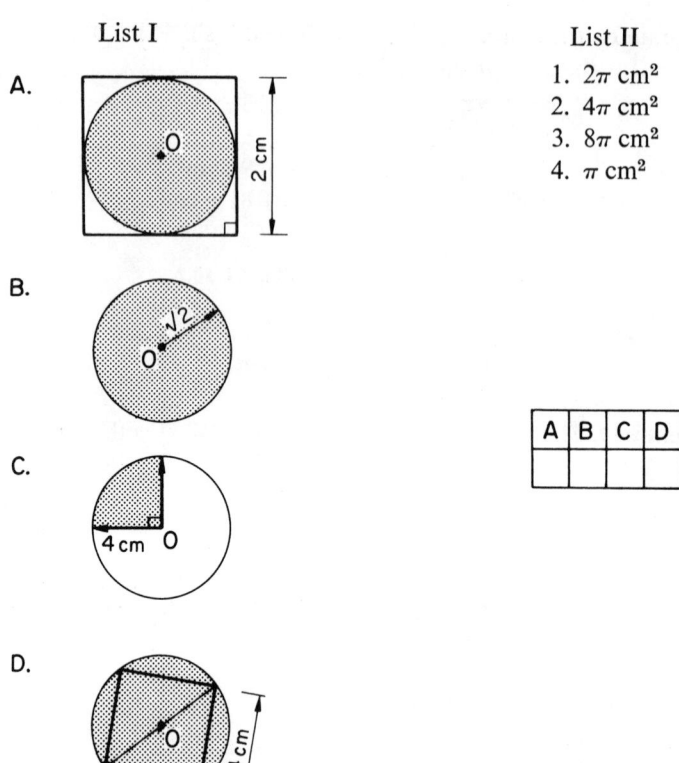

A.

B.

C.

D.

Fig. AT 10.5

List II

1. 2π cm²
2. 4π cm²
3. 8π cm²
4. π cm²

A	B	C	D

6. If the area of the parallelogram in the figure is 1 m² what is the area of the triangle?

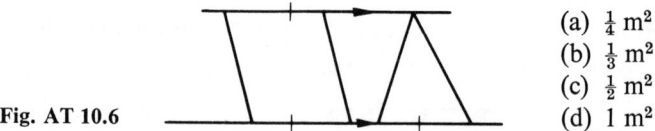

Fig. AT 10.6

(a) $\frac{1}{4}$ m²
(b) $\frac{1}{3}$ m²
(c) $\frac{1}{2}$ m²
(d) 1 m²

7. The area of the annulus shown in the diagram is

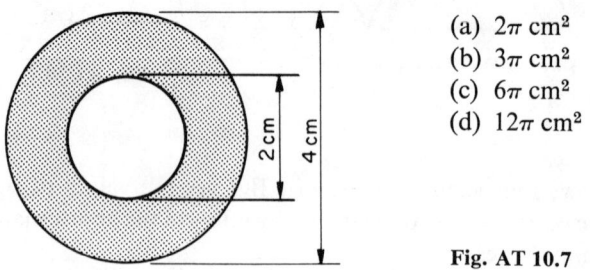

(a) 2π cm²
(b) 3π cm²
(c) 6π cm²
(d) 12π cm²

Fig. AT 10.7

8. List I consists of four figures, cube, sphere, cylinder, and cone. List II contains the expressions for the volume of each solid. Match the correct expression in List II to the figure in List I, by filling in the appropriate numbers in the vacant boxes.

List I

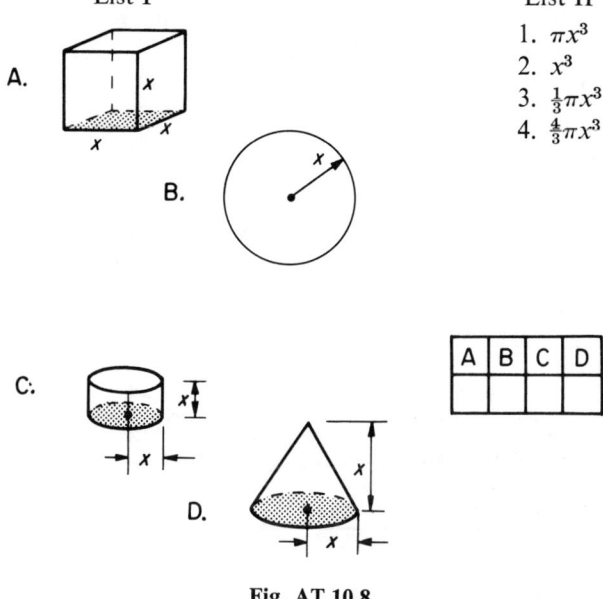

List II

1. πx^3
2. x^3
3. $\frac{1}{3}\pi x^3$
4. $\frac{4}{3}\pi x^3$

A	B	C	D

Fig. AT 10.8

9. All the solid figures in the diagram have the same formula for volume, in terms of the base area A and height h. Select the correct formula from the list.

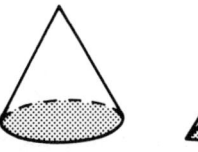 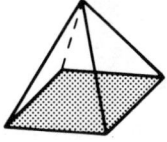 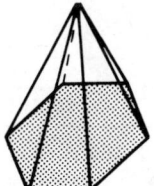

(a) $V = Ah$
(b) $V = \frac{1}{3}Ah$
(c) $V = \frac{1}{3}A^2 h$
(d) $V = A^2 h$

Fig. AT 10.9

10. A square-based prism has a volume of 24 cm³ and a height of 8 cm. What is the length of the side of the base?

   (a) 12 cm
   (b) √8 cm
   (c) 3 cm
   (d) √3 cm

11. The volume of a cylinder, of height 1 m is 3π m³. What is the height of a cone of the same base area and volume?

   (a) 3π m
   (b) π m
   (c) $\frac{1}{3}$ m
   (d) 3 m

12. The shaded plane figures are rotated one complete revolution about the axis XX. Name the solid figures so formed.

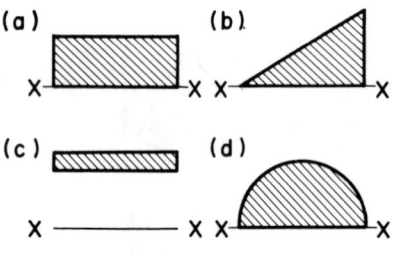

Fig. AT 10.10

13. A cube has a side of 1 cm. A cylinder has a radius of 1 cm and height 1 cm. In each case find

   (a) the volume of the solid
   (b) the total surface area of the solid
   (c) the ratio $\dfrac{\text{volume}}{\text{total surface area}}$

14. A cone and a cylinder have the same base area and the same height. The ratio of their volumes is

   (a) 1 : 1
   (b) 1 : 2
   (c) 1 : 3
   (d) 1 : 4

15. Name the solids which constitute each of the following solid figures.

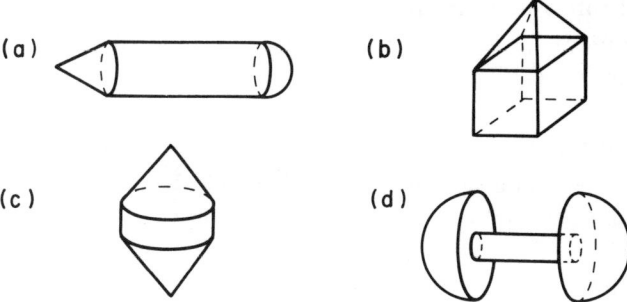

(a)   (b)

(c)   (d)

Fig. AT 10.11

16. The figure in the diagram is made up of a cylinder and cone, each of height x, and radius x. Find an expression for the volume of the figure in its simplest form.

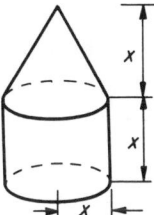

x

x

x

Fig. AT 10.12

What other figure will have the same expression for its volume?

17. If the radius of a cylinder is doubled, then its curved surface area will be multiplied by
   (a) 1
   (b) 2
   (c) 4
   (b) 8

18. The volume of a cylinder will be halved if
   (a) the radius is halved
   (b) the length is halved
   (c) the radius and length is halved
   (d) the radius is halved and the length is doubled.
   Choose the correct statement. More than one of them may be correct.

19. The area of a circle of radius 1 m is equal to the area of a semicircle of radius 2 m. Is this **true** or **false**?

20. A solid with a uniform cross-section is called a
    (a) pyramid
    (b) cone
    (c) prism
    (d) sphere

21. The perimeter of a semicircle of radius 1 m is
    (a) $2\pi + 2$
    (b) $\pi + 1$
    (c) $2\pi + 1$
    (d) $\pi + 2$

11. Trigonometry

Objectives

After working through this chapter you should be able to

1. Identify the names of the sides of a right-angled triangle in relation to a particular angle.
2. Sketch right-angled triangles given three suitable facts.
3. Write down the sine, cosine and tangent ratios for acute angles.
4. Calculate these trigonometrical ratios for any acute angle by drawing right-angled triangles.
5. Use four-figure trigonometrical tables to find the ratios for given acute angles.
6. Use these tables to find an acute angle given its sine, cosine or tangent ratio.
7. Calculate the sine, cosine and tangent ratios in fractional or root form.
8. Write down the relationship between sines and cosines of complementary angles.
9. Solve (that is, calculate unknown sides and angles) right-angled triangles using trigonometrical ratios and/or Pythagoras' theorem.
10. Solve practical problems involving right-angled triangles.
11. Draw graphs for sine and cosine in the range 0 to 90°.
12. Draw sine and cosine waves for angles in the range 0° to 360° by using the projections of a rotating radius of unit length.

11.1 Introduction

In Chapter 8 it was seen that Pythagoras' theorem could be used to evaluate the third side of a right-angled triangle provided the other two sides were known. It does not, however, give us a method of calculating the angles of the triangle.

Also, if an angle and one side of a right-angled triangle are known, geometrical methods do not enable us to calculate the remaining angles and sides. Trigonometry deals with the ratios between the sides of a right-angled triangle and it provides a method of calculating unknown sides and angles.

There are six possible ratios—known as the **trigonometrical ratios**—between the sides of a right-angled triangle but this chapter only deals with three of them, namely **sine, cosine** and **tangent**. It must be noted that the magnitudes of these ratios are independent of the scale of the right-angled triangle and depend only on the angles. For example, in Fig. 11.1, the ratios between the sides of triangles ABC are equal to the ratios between the sides of ADE, since these triangles are similar and their magnitude depends only on the value of the angle θ.

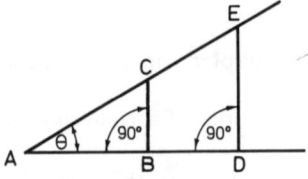

Fig. 11.1

Since the trigonometrical ratios depend upon the selection of pairs of sides it is necessary to be able to identify three sides. Fig. 11.2 shows how the sides are named in relation to a particular angle θ.

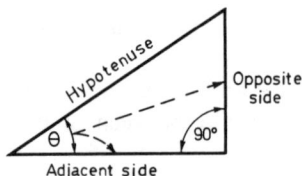

Fig. 11.2

11.2 Trigonometrical ratios of acute angles

These ratios are defined as follows:

The sine of angle θ

$$= \frac{\text{length of opposite side}}{\text{length of hypotenuse}} \quad \text{or} \quad \sin \theta = \frac{\text{opp. side}}{\text{hyp.}}$$

cosine of angle θ

$$= \frac{\text{length of adjacent side}}{\text{length of hypotenuse}} \quad \text{or} \quad \cos \theta = \frac{\text{adj. side}}{\text{hyp.}}$$

tangent of angle θ

$$= \frac{\text{length of opposite side}}{\text{length of adjacent side}} \quad \text{or} \quad \tan \theta = \frac{\text{opp. side}}{\text{adj. side}}$$

It is possible to find the trigonometrical ratios for any acute angle by drawing a right-angled triangle containing the required angle and hence finding the ratio of the sides.

242

EXAMPLE 11.1   Find the three trigonometrical ratios for an angle of 40°.
A right-angled triangle is first constructed containing an angle of 40°. Since the trigonometric ratios are independent of the scale of the triangle, a convenient length of 10·00 cm is used for BC (Fig. 11.3). The lengths of AB and AC are then measured from the completed triangle.

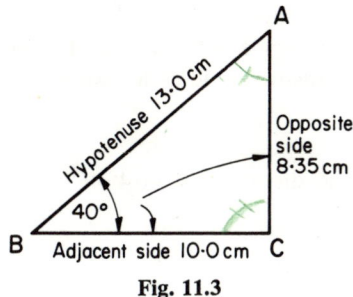

Fig. 11.3

The lengths of the sides are marked on the diagram. From the above definitions

$$\sin 40° = \frac{\text{opp. side}}{\text{hyp.}} = \frac{8·35}{13·00} = 0·642$$

$$\cos 40° = \frac{\text{adj. side}}{\text{hyp.}} = \frac{10·00}{13·00} = 0·769$$

$$\tan 40° = \frac{\text{opp. side}}{\text{adj. side}} = \frac{8·35}{10·00} = 0·835$$

Due to errors in measuring lengths the third significant figures in the above ratios may not be correct.

EXERCISE 11.1

Draw suitable right-angled triangles to determine as accurately as possible,
1. sin 50°   2. cos 70°   3. tan 36°

11.3   Use of trigonometrical tables

To avoid the tedious work of drawing triangles to calculate every trigono-metrical ratio a set of trigonometric tables can be used instead. In addition, the tables are accurate to four significant figures as opposed to two, or at the most three significant figures by drawing.
The tables give the value of the sine, cosine, and tangent of any acute angle.

The tables can therefore be used in two ways:

(a) knowing the angle, the ratios can be read from the tables

(b) knowing the ratio, the angle can be found.

Both methods are shown in the examples below but it must be noticed that whereas the sine and tangent of an angle increase as the angle increases, the cosine of an angle decreases as the angle increases. This means that the Mean Difference (M.D.) column must be

(a) *added* for increasing angle with sine and tangent

(b) *subtracted* for increasing angle with cosine.

EXAMPLE 11.2   Find the value of sin 4° 46′.

NATURAL SINES

Degrees	0′ 0°.0	6′ 0°.1	12′ 0°.2	18′ 0°.3	24′ 0°.4	30′ 0°.5	36′ 0°.6	42′ 0°.7	48′ 0°.8	54′ 0°.9	Mean Differences		
											1 2 3	4	5
0	0000	0017	0035	0052	0070	0087	0105	0122	0140	0157	3 6 9	12	15
1	0175	0192	0209	0227	0244	0262	0279	0297	0314	0332	3 6 9	12	15
2	0349	0366	0384	0401	0419	0436	0454	0471	0488	0506	3 6 9	12	15
3	0523	0541	0558	0576	0593	0610	0628	0645	0663	0680	3 6 9	12	15
4	0698	0715	0732	0750	0767	0785	0802	0819	0837	0854	3 6 9	12	15
5	0872	0889	0906	0924	0941	0958	0976	0993	1011	1028	3 6 9	12	14
6	1045	1063	1080	1097	1115	1132	1149	1167	1184	1201	3 6 9	12	14

Fig. 11.4

In Fig. 11.4

$$add \quad \sin \begin{cases} 4° \ 42′ = 0·0819 \\ \quad \ 4′ \qquad \quad 12 \ (M.D.) \end{cases}$$

$$\sin \ 4° \ 46′ = 0·0831$$

EXAMPLE 11.3   If sin θ = 0·0422, find the value of the angle θ.
In Fig. 11.4

$$(M.D.) \begin{matrix} 0·0419 = \\ \quad \ 3 \end{matrix} \ \sin \begin{cases} 2° \ 24′ \\ \quad 1′ \ add \end{cases}$$

$$0·0422 = \sin \ 2° \ 25′$$

so that

$$\theta = 2° \ 25′$$

EXAMPLE 11.4   Find the value of cos 45° 23'.

NATURAL COSINES

(Number in difference columns to be subtracted, not added.)

Degrees	0' 0°.0	6' 0°.1	12' 0°.2	18' 0°.3	24' 0°.4	30' 0°.5	36' 0°.6	42' 0°.7	48' 0°.0	54' 0°.9	Mean Differences		
											1 2 3	4	5
45	7071	7059	7040	7034	7022	7009	6997	6984	6972	6959	2 4 6	8	10
46	6947	6934	6921	6909	6896	6884	6871	6858	6845	6833	2 4 6	8 11	
47	6820	6807	6794	6782	6769	6756	6743	6730	6717	6704	2 4 6	9 11	
48	6691	6678	6665	6652	6639	6626	6613	6600	6587	6574	2 4 7	9 11	
49	6561	6547	6534	6521	6508	6494	6481	6468	6455	6441	2 4 7	9 11	
50	6428	6414	6401	6388	6374	6361	6347	6334	6320	6307	2 4 7	9 11	
51	6293	6280	6266	6252	6239	6225	6211	6198	6184	6170	2 5 7	9 11	

Fig. 11.5

In Fig. 11.5

$$add \quad \cos\begin{cases}45°\ 18' = 0\cdot7034 \\ 5' \qquad\quad 10\ (\text{M.D.}) \quad subtract.\end{cases}$$

$$\cos\ 45°\ 23' \quad 0\cdot7024$$

EXAMPLE 11.5   If cos θ = 0·6477, find θ.
In Fig. 11.5

$$0\cdot6481 = \cos\begin{cases}49°\ 36' \\ \end{cases}$$
$$subtract\ (\text{M.D.}) \quad 4 \qquad\qquad 2'\ add$$
$$0\cdot6477 = \cos\ 49°\ 38'$$

so that

$$\theta = 49°\ 38'$$

EXERCISE 11.2

1. Use trigonometrical tables to find the sine, cosine, and tangent of the following angles:
   (a) 10°   (b) 23° 2'   (c) 17° 11'
   (d) 19° 51'   (e) 62° 49'   (f) 85° 39'
2. Find the angles which have a sine of:
   (a) 0·9600   (b) 0·0092
   (c) 0·7005   (d) 0·9941
3. Find the angles which have a cosine of:
   (a) 0·4534   (b) 0·7090
   (c) 0·9340   (d) 0·0210
4. Find the angles which have a tangent of:
   (a) 0·4594   (b) 1·3769
   (c) 2·1900   (d) 6·3649

5. Evaluate:
   (a) $3\tan 27° + 2·4\tan 61°$
   (b) $\frac{1}{2}(\cos 14° \ 27' - \cos 55° \ 19')$
   (c) $\dfrac{\sin 27° + 3\sin 19°}{4\tan 24°}$
6. If $A = 41° \ 27'$ and $B = 22° \ 21'$ find the value of the expression

$$\frac{\tan A + \tan B}{1 - \tan A \tan B}.$$

11.4   Trigonometrical ratios for 30°, 45°, and 60°

The trigonometric ratio for these angles can be obtained quite easily in fractional or root form by considering suitable right-angled triangles. In some calculations it is convenient to use the ratios in this form.

(a) 45°

Consider the right-angled triangle ABC with $b = 1$ and $a = 1$ (Fig. 11.6).

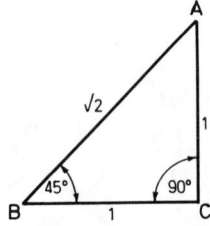

Fig. 11.6

The angles at A and B will be 45°. By Pythagoras

$$c^2 = a^2 + b^2$$

$$= 1^2 + 1^2 = 2$$

$$\therefore \quad c = \sqrt{2}$$

Now consider angle B as θ

$$\sin \theta = \frac{\text{opp. side}}{\text{hyp.}} \qquad \therefore \quad \sin 45° = \frac{1}{\sqrt{2}} = 0·7071$$

$$\cos \theta = \frac{\text{adj. side}}{\text{hyp.}} \qquad \therefore \quad \cos 45° = \frac{1}{\sqrt{2}} = 0·7071$$

$$\tan \theta = \frac{\text{opp. side}}{\text{adj. side}} \qquad \therefore \quad \tan 45° = \frac{1}{1} = 1·0000$$

246

(b) 30° and 60°

Consider the equilateral triangle ABD with sides of length 2. (Fig. 11.7). AC bisects the vertical angle. By symmetry it will also bisect the base at 90°.

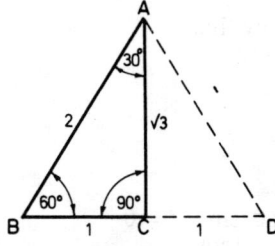

Fig. 11.7

In $\triangle$ ABC, by Pythagoras,

$$c^2 = a^2 + b^2$$
$$\therefore \ b^2 = c^2 - a^2$$
$$= 2^2 - 1^2 = 3$$
$$\therefore \ b = \sqrt{3}$$

Taking $\angle$ ABC as θ, then

$$\sin 60° = \frac{\sqrt{3}}{2} = 0\cdot 8660$$

$$\cos 60° = \tfrac{1}{2} \ = 0\cdot 5000$$

$$\tan 60° = \frac{\sqrt{3}}{1} = 1\cdot 7321$$

Taking $\angle$ BAC as θ, then

$$\sin 30° = \tfrac{1}{2} \ = 0\cdot 5000$$

$$\cos 30° = \frac{\sqrt{3}}{2} = 0\cdot 8660$$

$$\tan 30° = \frac{1}{\sqrt{3}} = 0\cdot 5774$$

EXERCISE 11.3

1. Without using tables, evaluate the following expressions:

(a) $2\cos 30° + 4\sin 60°$　　(b) $\dfrac{\tan 60°}{\tan 30°}$

(c) $2 \sin 45° + 3 \cos 45°$ (d) $\dfrac{3 \cos 60°}{2 \sin 60°}$

The answers can, if necessary, be left in fractional or root form.

2. Without using tables show that
   (a) $\sin 60° = 2 \sin 30° \cos 30°$
   (b) $\sin^2 45° + \cos^2 45° = 1$

   (c) $\dfrac{\sin 30°}{\cos 30°} = \tan 30°$

   (d) $\cos 60° \cos 30° - \sin 60° \sin 30° = 0$
   (e) $\cos 60° = \cos^2 30° - \sin^2 30°$

11.5   Sine and cosine of complementary angles

Two angles are complementary if they add up to 90°.
   It will be noticed that
$$\sin 60° = \cos 30°$$
and
$$\sin 30° = \cos 60°$$

This will be true for any pair of complementary angles, for example
$$\sin 47° = \cos 43°, \quad \cos 24° = \sin 66°$$

that is, the sine of any acute angle is equal to the cosine of its complement, and the cosine of any acute angle is equal to the sine of its complement. This statement can be summarized as follows.
   For any angle θ its complement is $(90 - \theta)$. Therefore
$$\sin \theta = \cos (90 - \theta)$$
$$\cos \theta = \sin (90 - \theta)$$

11.6   Solution of right-angles triangles

To solve right-angled triangles (i.e., to find any unknown sides and angles) by trigonometrical methods either (a) two sides, or (b) one side and one angle, must be known. Three trigonometrical formulae are available and the student must select the one containing the two facts that are known and one fact that is to be found.

EXAMPLE 11.6 Calculate angle A in the right-angled triangle ABC if $a = 3{\cdot}8$ m and $b = 5{\cdot}3$ m.

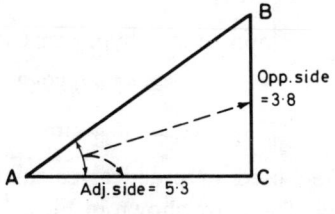

Fig. 11.8

Since the opposite and adjacent sides to angle A are known then the tangent ratio must be used, i.e.,

$$\tan A = \frac{3{\cdot}8}{5{\cdot}3} = 0{\cdot}7169$$

Therefore from tangent tables, angle $A = 35° 38'$.

EXAMPLE 11.7 In right-angled triangle ABC, if angle $B = 64°$ and $AB = 65$ mm, find the lengths of the sides BC and AC.

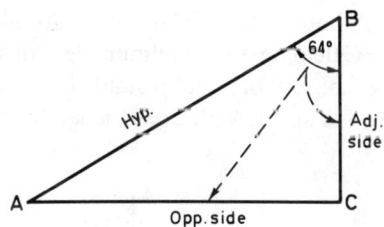

Fig. 11.9

$$\frac{\text{adj. side}}{\text{hyp.}} = \cos 64°$$

$$\therefore \quad \text{adj. side} = \text{hyp.} \times \cos 64°$$

$$= 65 \times 0{\cdot}4384$$

$$BC = 28{\cdot}5 \text{ mm}$$

AC can now be found by Pythagoras or a trigonometrical method. The latter method is used here

$$\frac{\text{opp. side}}{\text{hyp.}} = \sin 64°$$

$$\therefore \quad \text{opp. side} = \text{hyp.} \times \sin 64°$$

$$= 65 \times 0.8988$$

$$AC = 58.4 \text{ mm}$$

Two expressions often used in practical calculations are the angles of **elevation** and **depression**. These are shown in Figs. 11.10 and 11.11.

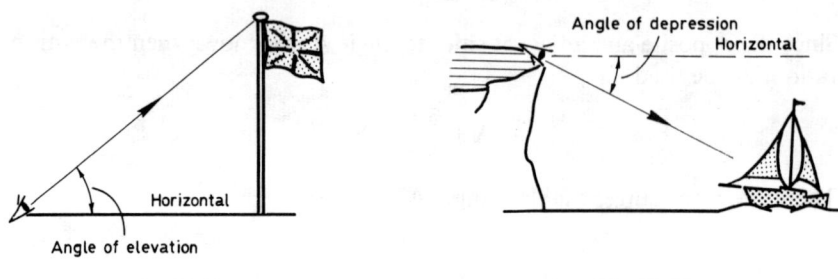

Fig. 11.10   Fig. 11.11

EXAMPLE 11.8   A house has a flat-roofed outbuilding 2 m wide by 2 m high attached to it. Calculate the minimum length of ladder required to reach the wall of the house from the ground if the ladder has an angle of elevation of 60°. How far up the wall does the ladder reach?

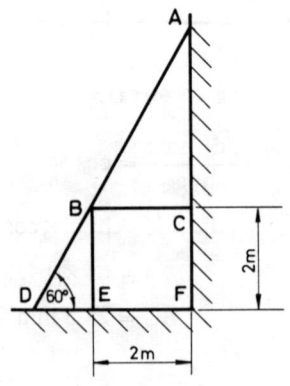

Fig. 11.12

250

In △ ABC

$$\cos 60 = \frac{BC}{AB} = \frac{2}{AB}$$

$$\therefore \quad AB = \frac{2}{\cos 60} = \frac{2}{0 \cdot 5}$$

$$= 4 \text{ m}$$

In △ BED

$$\sin 60 = \frac{BE}{BD} = \frac{2}{BD}$$

$$\therefore \quad BD = \frac{2}{\sin 60} = \frac{2}{0 \cdot 866}$$

$$= 2 \cdot 3 \text{ m}$$

Length of ladder $= 4 + 2 \cdot 31$

$$= 6 \cdot 3 \text{ m}$$

In △ ADF

$$\sin 60 = \frac{AF}{AD} = \frac{AF}{6 \cdot 31}$$

$$\therefore \quad AF = 6 \cdot 31 \times \sin 60$$

$$= 6 \cdot 31 \times 0 \cdot 866$$

Distance up wall $= 5 \cdot 464 \text{ m}$

$$= 5 \cdot 5 \text{ m}$$

EXERCISE 11.4

1. ABC is a right-angled triangle with angle B = 90°.
   (a) If $a = 4 \cdot 8$ and $b = 7 \cdot 6$ find angle A.
   (b) If $a = 6 \cdot 7$ and $b = 9 \cdot 4$ find angle C.
   (c) If $a = 2 \cdot 9$ and $c = 1 \cdot 9$ find angle A.
2. In a right-angled triangle the side adjacent to one of the acute angles is 16·9 mm and the side opposite this angle is 16·2 mm. What are the angles of the triangle?
3. ABC is a right-angled triangle.
   (a) If $\angle$ B = 90°, $\angle$ A = 40°, and $a = 24 \cdot 3$ find sides b and c.
   (b) If $\angle$ A = 90°, $\angle$ B = 74°29′, and $c = 1 \cdot 48$ find the other sides.
   (c) If $\angle$ A = 90°, $\angle$ C = 62° 16′, and $a = 3 \cdot 59$ find the other sides.
4. In a right-angled triangle the hypotenuse is 6·33 m long and one of the angles is 33° 30′. What is the length of the side opposite the angle?

5. ABC is a right-angled triangle with angle A = 90° and AD is a line perpendicular to BC. If ∠ ACB = 35° and AB = 12 mm, calculate the lengths of AC and DC.

6. In triangle ABC angle B = 90° and a perpendicular is drawn from B to AC meeting it at D. If ∠ C = 64° and AB = 72 mm find the length of AD.

7. An isosceles triangle has equal sides of length 45 m and a base length of 32 m. What are the angles of the triangle?

8. A rhombus has sides of length 40 mm. If the angle between two of the sides is 62°, find the lengths of the diagonals.

9. In a parallelogram ABCD the diagonal AC is perpendicular to side BC. If BC = 70 mm and ∠ BAC = 32° calculate the length of side AB.

10. ABC is an isosceles triangle with AB = AC = 142 mm and angle ABC is 58°. D is a point on AB such that the line CD is perpendicular to AB. Find the length of CD.

11. If a ladder of length 7 m is inclined at 63° to the horizontal, how far up the wall will it reach?

12. A cone has a semi-vertical angle of 20°. If the slant height is 24 mm what is the base diameter?

13. One end of a tie wire, 6 m long, is tied to the top of a vertical post, 4·2 m high, and the other end is fixed to the ground. What is the inclination of the wire to the horizontal?

14. Calculate the vertical height of a tower if the angle of elevation of the top of the tower when viewed from a point 20 m horizontally from the tower base is 64° 12′.

15. The angle of depression of a yacht when observed from the top of a cliff 0·32 km high is 12°. What is the distance of the yacht from the base of the cliff?

16. A straight girder rises 0·1 m every 3 m of its length. What is the inclination of the girder to the horizontal?

17. A telephone post is 9 m from a house, measured horizontally. Calculate the length of cable required to pass from the top of this post to a point on the wall of the house if the wire is inclined at 27° to the horizontal.

18. In a reciprocating engine the crank AB = 0·3 m long and the con-rod BC is 1·25 m long. If the angle between the crank and the con-rod is 90° determine the distance AC and the angle BAC.

19. A pendulum of length 600 mm swings through an angle of 12° each side of the vertical. Calculate the vertical movement of the lower end of the pendulum.

20. Determine the dimension x for the sheet of aluminium shown in Fig. 11.13.

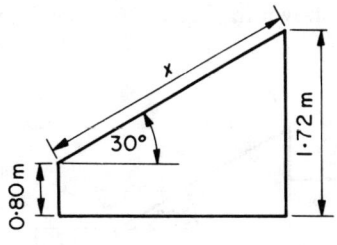

Fig. 11.13

21. Fig. 11.14 shows the end-view of a symmetrical building. Calculate the total sloping length of the roof. Find, also, the pitch (i.e., the inclination to the horizontal) of the roof.

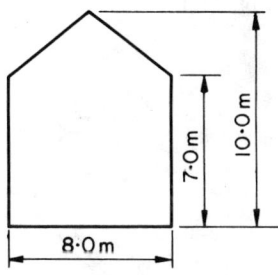

Fig. 11.14

22. Find the angle θ for the tapered roller shown in Fig. 11.15.

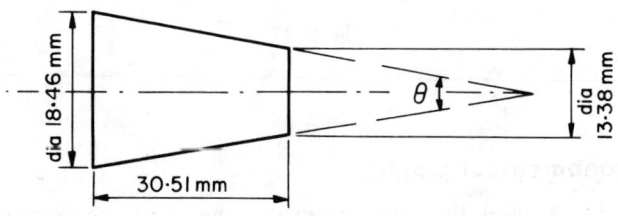

Fig. 11.15

23. A tool has a shape shown in Fig. 11.16. Calculate the lengths x and y.

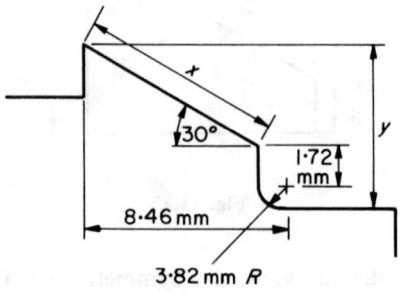

Fig. 11.16

24. Calculate the distances x and y of the centres of the holes from the left-hand datum face AB (Fig. 11.17).

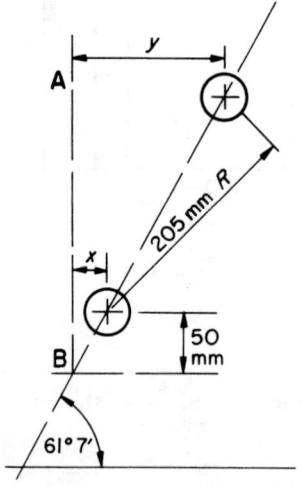

Fig. 11.17

11.7 Trigonometrical graphs

The graphs below show the variation in the sine and cosine ratios over the angle range 0° to 90°. In each case the angle scale is plotted horizontally and the ratios, obtained from the tables for each angle, are plotted vertically.

(a) Graph of sin θ

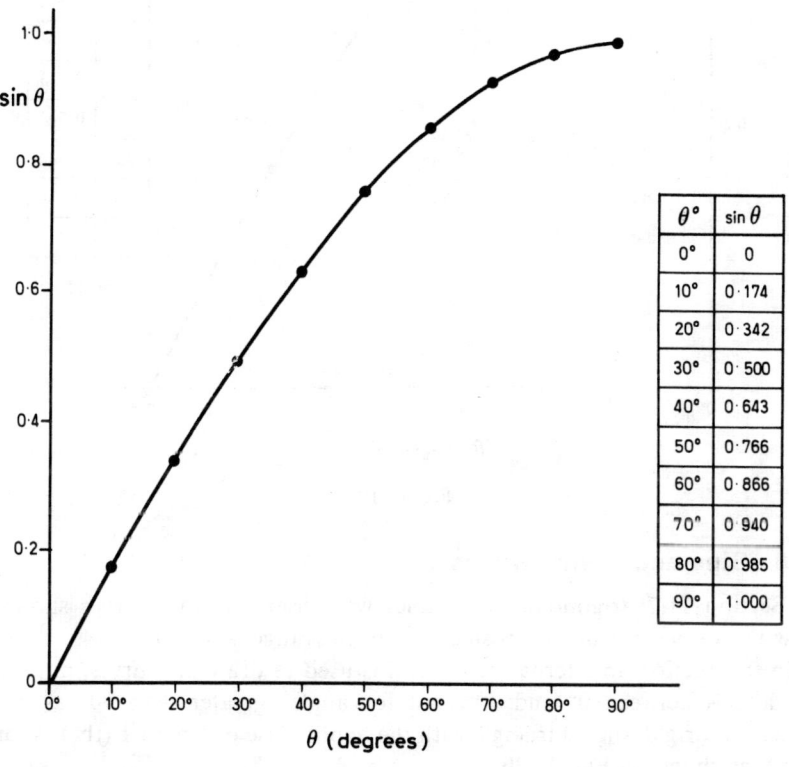

$\theta°$	$\sin\theta$
0°	0
10°	0·174
20°	0·342
30°	0·500
40°	0·643
50°	0·766
60°	0·866
70°	0·940
80°	0·985
90°	1·000

Fig. 11.18

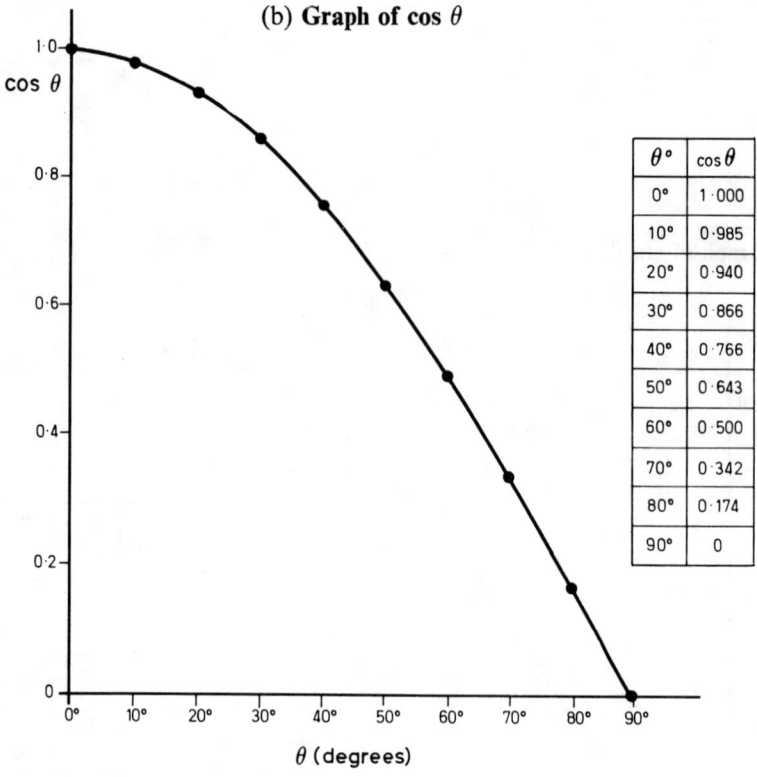

(b) Graph of cos θ

$\theta°$	$\cos\theta$
0°	1·000
10°	0·985
20°	0·940
30°	0·866
40°	0·766
50°	0·643
60°	0·500
70°	0·342
80°	0·174
90°	0

θ (degrees)

Fig. 11.19

11.8 Sine and cosine waves

In Section 11.7, trigonometrical tables were used to draw curves showing how the values of sine and cosine vary in the range 0° to 90°.

In this section an alternative method is used to draw the curves, not only for angles between 0° and 90°, but for angles greater than 90° as well. Consider a right-angled triangle with the hypotenuse of length 1 (that is, one unit), as shown in Fig. 11.20.

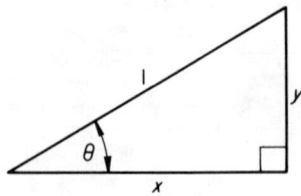

Fig. 11.20

Now

$$\sin \theta = \frac{y}{1}, \quad \text{so that} \quad y = \sin \theta$$

$$\cos \theta = \frac{x}{1}, \quad \text{so that} \quad x = \cos \theta$$

It is seen, therefore, that the vertical distance y gives the sine of the angle θ, and the horizontal distance gives the cosine of the angle θ. x and y are called the horizontal and vertical projections of the hypotenuse.

To draw the sine and cosine curves the hypotenuse of length 1 (that is, one unit) is rotated anticlockwise and the projections are plotted as shown in Fig. 11.21.

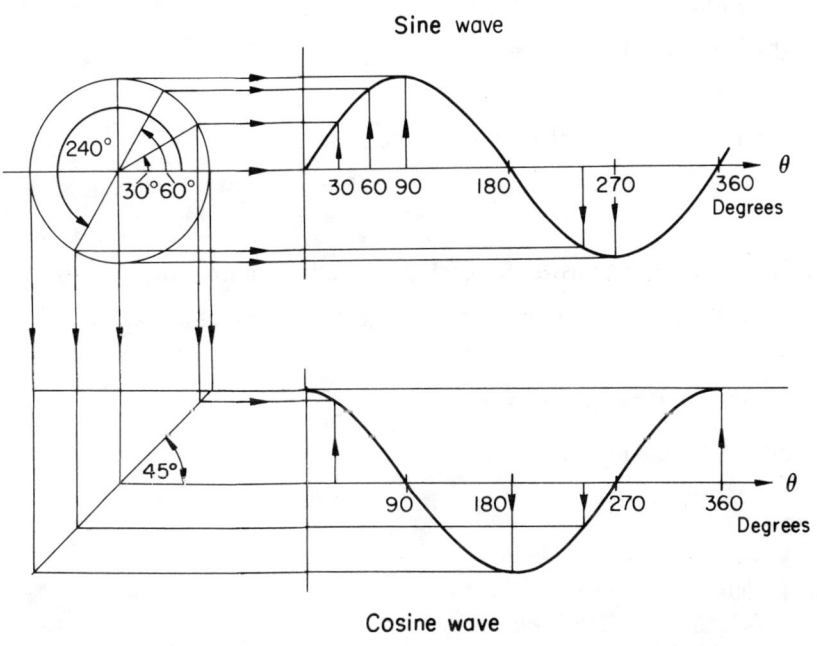

Fig. 11.21

If these curves are examined the following features may be observed.
(a) A cosine curve is identical to a sine curve except that it is shifted 90° to the left of the sine curve. This is to be expected because as seen in Section 11.5, for complementary angles $\sin \theta = \cos(90 - \theta)$.
(b) After 360° both curves repeat themselves, that is, one complete cycle of the waves has occurred in the range 0 to 360°.

Assessment test 11

1. State whether the following are **true** or **false**:
    (a) If $\cos \theta = 0$, then $\theta = 0°$.
    (b) The maximum value of $\sin \theta$ is 1.
    (c) $\cos 45 = \sqrt{2}$.
    (d) The sine of $0°$ is 0.
    (e) $\sin 45° = \cos 45°$

2. Fill in the blanks in each of the following, without using trigonometrical tables:
    (a) $\sin 45° = $ ....
    (b) If $\tan \theta = 1$, then $\theta = $ ....
    (c) $\cos$............ $= \frac{1}{2}$.
    (d) If $\sin \theta = \cos \theta$, then $\theta = $ ....

3. Select the correct answer for each of the following from the list 1–6:
    (a) If the angles α and β are complementary, then $\alpha + \beta = $ ....
    (b) $\sin 17° = \cos$ ....
    (c) If $\cos 27° 18' = 0.8886$, then $\sin 62° 42' = $ ....
    (d) $\sin \theta = \cos ($............$)$.
    1. $73°$
    2. $(90 - \theta)$
    3. 90
    4. $(90 + \theta)$
    5. 0.8886
    6. $1 - 0.8886$

4. *Sketch* the following right-angled triangles from the data given:
    (a) hypotenuse $= 10$ cm, one angle $= 40°$
    (b) one angle $= 75°$, side opposite this angle $= 6$ cm
    (c) one angle $= 50°$, side adjacent to this angle $= 5$ cm
    (d) $\angle C = 90°$, $b = 5$ cm, $c = 8$ cm

258

5. List I contains four right-angled triangles showing the angle θ. List II contains four values of sin θ. Match the correct ratio to each diagram by filling in the appropriate numbers in the vacant boxes.

List I   List II

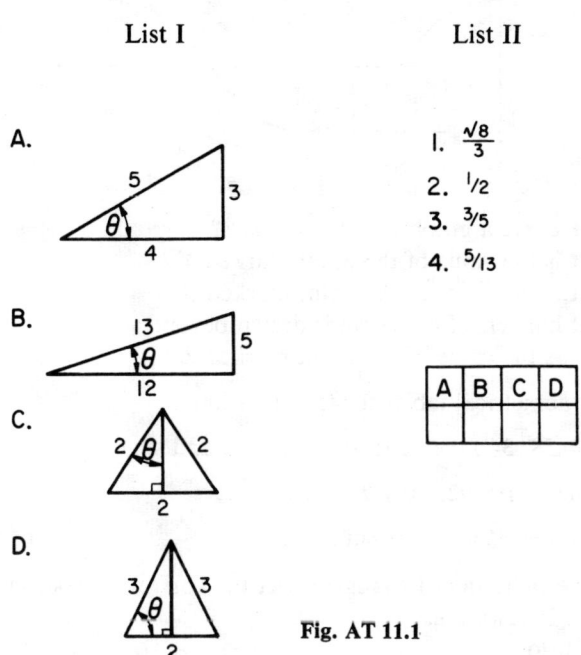

A.

B.

C.

D.

1. $\frac{\sqrt{8}}{3}$
2. $^1/_2$
3. $^3/_5$
4. $^5/_{13}$

A	B	C	D

Fig. AT 11.1

6. With reference to List I in question 5, List II contains four values of cos θ. Match the correct value to the triangles in List I in question 5 by filling in the appropriate numbers in the vacant boxes.

List II

1. $\frac{\sqrt{3}}{2}$
2. $\frac{1}{3}$
3. $\frac{12}{13}$
4. $\frac{4}{5}$

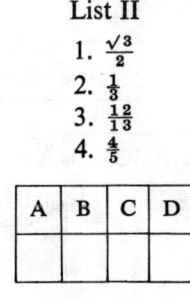

A	B	C	D

Fig. AT 11.2

7. With reference to the graph shown in the diagram answer the following questions.

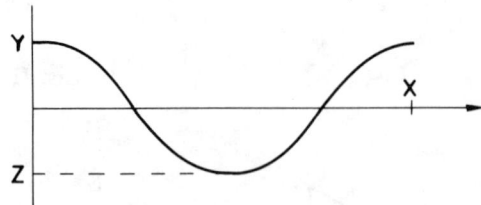

Fig. AT 11.3

(a) Is the curve a graph of sin θ or cos θ?
(b) What is the value of the point marked Y?
(c) What is the value of the point marked X?
(d) What happens if the curve is drawn beyond X?
(e) What is the value of the point marked Z?

8. Use trigonometrical tables for this question.

(a) If $\theta = 38° 34'$, then cos $\theta = \ldots\ldots\ldots$ and sin $\theta = \ldots\ldots\ldots$.

(b) If tan $\theta = 1\cdot1792$, then $\theta = \ldots\ldots\ldots$.

(c) If cos $\theta = 0\cdot5668$, then sin $\theta \ldots\ldots\ldots$.

9. From the right-angled triangle select the correct expressions for b and c from the following list.

(a) $10\sin 40$
(b) $10\cos 40$
(c) $10\tan 40$
(d) $10/\sin 40$

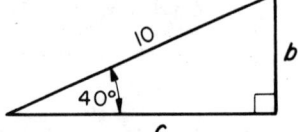

Fig. AT 11.4

10. From the right-angled triangle select the correct values for b and c from the following list.

(a) $\frac{\sqrt{3}}{2}$
(b) 1
(c) $\frac{1}{\sqrt{2}}$
(d) $\sqrt{3}$

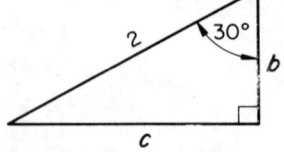

Fig. AT 11.5

11. In the right-angled triangle ABC, $\cos A = \frac{12}{13}$. Which of the following statements are correct?

(a) $\sin C = \cos A$
(b) $\cos B = \frac{5}{12}$
(c) $\tan C = 2\cdot2$
(d) $\tan A = 0\cdot4167$

Fig. AT 11.6

260

12. If $\tan \theta = \frac{3}{4}$, which of the following triangles is correctly labelled?

(a)

(b)

(c)

(d)

Fig. AT 11.7

Revision exercise

Section A: Arithmetic

1. (a) A spring extends 2 cm under a load of 1·5 kN. Calculate (i) the extension for a load of 400 N, and (ii) the load required for an extension of 1·8 cm.

   (b) A gear wheel having 35 teeth revolves at 140 rev/min. It meshes with a wheel having 80 teeth. Find the speed of the 80-tooth wheel.

   (EMEU)

2. (a) Three resistors in the ratio 2 : 3 : 7 have a total resistance of 900 ohm. What is the resistance of each resistor?

   (b) A batch of 330 ohm resistors is marked with a tolerance of ±5%. What are the maximum and minimum resistance values in the batch?

   (c) A lathe operation is completed in 1 min 42s. How long would it take for the same operation if the time were reduced by 2%?

   (EMEU)

3. (a) A casting of mass 120 kg consists of copper, zinc, and manganese in the ratio 60 : 39 : 1 (by mass). Calculate the mass of **each** constituent in the casting.

   (b) In a factory three machines are used to produce a component. If each machine produces 6000 components, of which 7%, 3%, and 2% are faulty respectively, calculate the percentage of faulty items in the total output from the factory.

   (ULCI)

4. (a) A machine produces washers at the rate of 1000 per h for the first 3 h of a shift and 1400 per h for the next 5 h. What is the average hourly rate of production during the shift?

   (b) The proportion of men, women, and apprentices employed in a factory is 10 : 2·5 : 1·5. How many (i) women and (ii) apprentices are there if the total number of people employed is 1820?

5. (a) A wire 105 m long has a resistance of 8 ohm. Assuming resistance is directly proportional to length find (i) the resistance of a wire 75 m long, and (ii) the length of a wire which has a resistance of 11 ohm.

   (b) A test on a 0–50 V voltmeter shows that the voltmeter reads 23·5 V when the true value is 25 V. Determine the per unit error in the reading.

   (EMEU)

6. (a) Without using tables or aids to calculation find the value of

$$\frac{2 \cdot 4 \times 27}{0 \cdot 09 \times 0 \cdot 8}$$

giving the answer in standard form.

(b) Determine 7% of £251·6.

(c) Using log tables and showing all working calculate B, to **three** significant figures, when $A = 6 \cdot 875$ and $C = 197 \cdot 5$:

$$B = \frac{A + C}{2AC}.$$

(NCTEC)

7. (a) The average resistance of 25 resistors is 30 ohm. If the average resistance of 24 of these resistors is 28 ohm, determine the resistance of the remaining resistor.

(b) Evaluate using logarithms

$$\frac{7 \cdot 482 \times 48 \cdot 37}{0 \cdot 0793 \times 2 \cdot 550}$$

(c) Calculate the perimeter of a triangle ABC right angled at B, where $AC = 95$ mm and $BC = 36$ mm.

(UEI)

8. With the use of logarithms, and/or other tables, evaluate

$$\sqrt{\left(\frac{9 \cdot 81 + \sqrt{(32 \cdot 2)}}{0 \cdot 7854^2}\right)}$$

giving the answer correct to two decimal places.

(ULCI)

9. (a) Using square root and reciprocal tables **only** evaluate

$$\frac{2}{\sqrt{(7 \cdot 458)}}$$

(b) Re-write the following as indicated:
    (i) 23·98 to three significant figures
    (ii) 0·01656 to two significant figures
    (iii) 0·0364 to two decimal places.

(EMEJ)

10. (a) Find the value of $(1\frac{3}{4} + 4\frac{7}{16}) \times 1\frac{1}{2} : 1\frac{3}{8}$.

(b) Use appropriate tables to find the values of (i) $8 \cdot 339^2$, (ii) $\sqrt{(446 \cdot 7)}$.

(c) Express $\frac{3}{7}$ as a decimal giving your answer correct to **two** decimal places.

(d) Express 0·24 as a fraction in its lowest form.

(UEI)

11. (a) Calculate the values of

   (i) $\sqrt{[(8{\cdot}62)^2+(27{\cdot}4)^2]}$ (ii) $\left(\dfrac{1}{5{\cdot}556}\right)^2$

   (b) Without using tables evaluate

   $$\frac{7\times 10^3 \times 9 \times 10^3}{3\times 10^5}$$

   expressing the answer in standard form.

   (UEI)

12. The minimum radius which a vehicle can negotiate is given by the formula

   $$r=\frac{2v^2 h}{ag}$$

   Using logarithms, determine the value of r when $v = 35$, $h = 0{\cdot}6734$, $a = 1{\cdot}314$ and $g = 9{\cdot}81$.

   (EMEU)

13. (a) Use logarithmic tables to evaluate:

   (i) $\dfrac{17{\cdot}6-4{\cdot}1}{0{\cdot}291}$ (ii) $\dfrac{284}{7615\times 0{\cdot}0831}$

   (b) If $v^2-u^2 = 2ax$ calculate by any method v when $u = 4{\cdot}7$, $a = 9{\cdot}8$, and $x = 1{\cdot}5$.

   (EMEU)

14. (a) In making a television set the cost of labour and materials is in the ratio 5 : 4. The manufacturer sells the set for £84·00 and 25% of this is profit. What is the cost of the materials for the set?
   (b) Without using tables find the value of

   $$\frac{3{\cdot}6\times 48}{9\times 0{\cdot}6}$$

   giving the answer in standard form.

   (UEI)

15. (a) Use logarithmic tables to evaluate

   $$9{\cdot}761 - \left(\frac{0{\cdot}06\times 0{\cdot}561}{0{\cdot}004\ 31}\right)$$

   (b) If $R_1 = R_0[1+\alpha(t_2-t_1)]$ determine R_1 when $R_0 = 21{\cdot}74$, $t_2 = 60{\cdot}4$, $t_1 = 18$ and $\alpha = 0{\cdot}0016$.

   (EMEU)

16. (a) Use logarithmic tables to find the value of

$$\frac{1}{376 \cdot 2}\left(\frac{6 \cdot 25}{0 \cdot 0719} - 0 \cdot 8734\right)$$

giving your answer correct to **two** decimal places.

(b) Using relevant four-figure tables (not logarithms), evaluate

$$\sqrt{(650 \cdot 8)} + \frac{1}{33 \cdot 24}$$

(ULCI)

Section B: Algebra

17. (a) Simplify:
   (i) $(\frac{1}{2}a^3 b)^3 \times 4(2ab^2)^2$

   (ii) $\dfrac{a^{-1}b^{-1}c^{-2} \times 14a^2 b^3 c}{7c^{-1}}$

(b) Expand
   (i) $(2a+b)(a-2b)$ (ii) $(a-b)^2$.

(EMEU)

18. (a) Simplify $(3xy^3 z^2)^4$.

(b) Write down an equation to represent the following statement:
   The sum of the squares of two numbers x and y is equal to three times the product of the two numbers.

(c) Solve the equation $3x - 5(2x+7) = 0$.

(NCTEC)

19. (a) Find the value of the expression

$$x^2 - 2xy + y^2$$

when

$$x = -2 \quad \text{and} \quad y = -3$$

(b) Transpose the formula

$$s = \left(\frac{u+v}{2}\right)t$$

to give u in terms of the other symbols.

(c) Solve the following equation for x

$$5x - 3(x-2) + 2(2x-1) = 13$$

(UEI)

20. (a) Solve
$$2(3m-6)-4(6m-4) = 3-(2m+1)$$

(b) Solve the simultaneous equations:
$$2 \cdot 5a - 3b = -0 \cdot 45$$
$$1 \cdot 6a + 0 \cdot 8b = 0 \cdot 8$$

(EMEU)

21. (a) Solve the equation $5(y-3)+43 = 3(4-y)$.
(b) Simplify $4[3(x-5)-3(2-2x)]$.
(c) Factorize
    (i) $2xy+4ax-6x$
    (ii) $3(x-2)+4y(2-x)$

(ULCI)

22. (a) Solve the equation $5(2y-7)-4(3y+1)+45 = 0$.
(b) When multiplied out $(2x-3)(4x-1)$ is to be the same as $8x^2+kx+3$. Find k.

(YHCFE)

23. Transpose the following formulae to make the letter in the square bracket the subject of the formulae:

(a) $R = \dfrac{\rho l}{a}$ [a]

(b) $V = \dfrac{\pi R^2 h}{3}$ [R]

(c) $V^2 = U^2 + 2as$ [a]

(d) $d = \dfrac{1}{S} + \dfrac{P}{20}$ [S]

(e) $\dfrac{P}{W} = \dfrac{t+u}{1-t}$ [t]

(EMEU)

24. (a) Simplify the following, writing the answers with positive indices only:

(i) $\dfrac{5a^2 b \times 2abc^2}{15a^3 c}$ (ii) $\dfrac{(3ab)^2 \times (ab^2)^3}{a^{-2}b^{-3}}$

(b) Solve the equation
$$\frac{24}{x} + 3 = 9$$

(EMEU)

266

25. (a) Simplify $8ay^2 \times 2a^2 y \div (4ay)^2$.
    (b) Multiply $(a+2b)$ by $(2a+b)$.
    (c) Simplify
        (i) $2a+b-a-2b$ (ii) $x+3y-(2y-x)$.
    (d) Solve the equation $3(x+2)+2(3x+4) = 41$.

(UEI)

26. (a) Solve the equation $3(4x-2)-2(3x+7) = 5(x+1)$.
    (b) Use the laws of indices to simplify:

    (i) $\sqrt{\left(\dfrac{25a^3 bc^2}{4ab^5}\right)}$ (ii) $\dfrac{10^3}{10^{-1} \times 10^5}$

    (c) Make s the subject of the following formulae:

    (i) $y = \dfrac{2s+y}{2x}$ (ii) $\dfrac{4}{p} = \dfrac{2}{s} + \dfrac{3}{r}$

(NCTEC)

27. (a) Solve the equations:

$$2k - 3t = 17$$

$$5k + 6t = 2$$

    (b) In finding the reactions at the supports of a beam, the following equations arose:

$$P + Q = 32$$

$$13P = 3Q$$

    Calculate P and Q.

(YHCFE)

Section C: Graphs

28. In an experiment the velocity v of a projected body was noted at various times t, providing the following data:

v	37·3	30·4	23·5	17·5	2·7
t	1·1	1·8	2·5	3·1	4·6

By plotting the values of t horizontally and the values of v vertically, show graphically that v and t are connected by a straight line.
    From the graph find the value of
    (a) v when $t = 3·75$ and
    (b) t when $v = 25$.

(YHCFE)

29. What is the gradient of the line AB in each of the cases shown in Fig. R 1.
(EMEU)

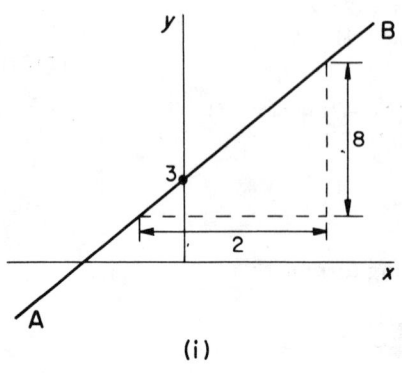 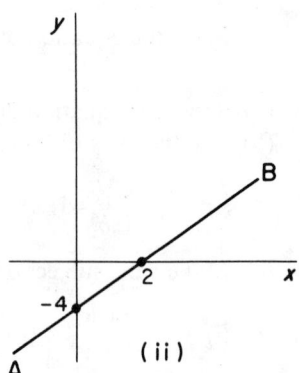

(i)   (ii)

Fig. R 1

30. A close-coiled helical spring was loaded in compression and the length of the spring L was measured when various forces F were applied. The following results were obtained:

Length L (mm)	63·5	61·0	58·5	56·0	53·5	51·0
Force F (N)	50	60	70	80	90	100

Plot these values and show that the graph is a straight line.
What is the nominal unloaded length of the spring?

(UEI)

31. The following table shows the length l millimetres of a helical spring when supporting a mass of m kilograms.

m (kg)	2	3·5	5	6	8	9	10
l (mm)	225	325	420	480	610	675	740

Plot a graph of mass (horizontal axis) against length of spring. Use the graph to determine
  (i) the length of the spring when no mass is attached
  (ii) the mass which would cause the spring to extend to a length of 450 mm
  (iii) the change of length per kilogram of mass added.

(UEI)

32. In the electric spot-welding process the maximum number of spot welds produced per minute is related to the thickness of the metal being welded.

Metal thickness (mm)	1·6	1·8	2·0	2·2	2·4	2·8
Spots per minute	10	15	33	42	62	80

Plot a graph, spots per minute vertically, draw the best straight line through the points and determine the number of spot welds per minute on 2·6 mm sheet.

(NCTEC)

33. In an experiment to determine the internal resistance of a lead–acid cell the following results of terminal potential difference V and current I were as follows:

I (A)	1·0	1·5	2·0	2·5	3·0	4·0	5·0
V (V)	2·05	1·90	1·75	1·63	1·48	1·20	0·93

Draw the graph of V (vertical axis) against I (horizontal axis). Show that the graph is a straight line.

Use this law to determine the value of I when $V = 0$.

(UEI)

Section D: Geometry

34. A flat is to be machined on a 100 mm diameter bar as shown. Calculate the dimension x.

(UEI)

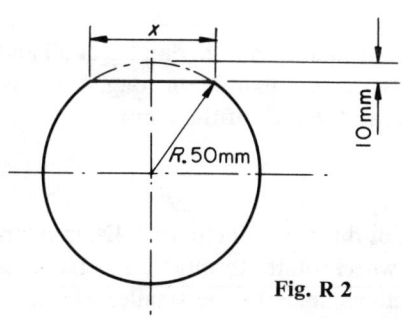

Fig. R 2

35. A metal template is in the form of an isosceles triangle of base 120 mm and height 180 mm. The top of the triangle is removed by a cut parallel to the base. If the length of the cut is 70 mm calculate the height of the remaining portion.

(UEI)

36. (a) The sides of a triangle are 2 cm, 5 cm, and 6 cm respectively. Calculate the length of the perimeter of a similar triangle the shortest side of which is 7 cm.

(b) Calculate the length *BD* shown in Fig. R 3.

(EMEU)

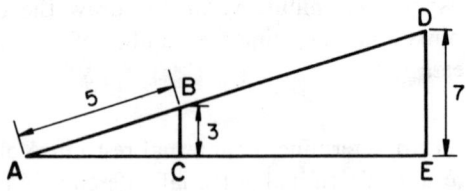

Fig. R 3

37. XYZ is a right-angled triangle with the right angle at Y. PY is a perpendicular drawn from Y to XZ.

Draw this figure and identify the three similar triangles. If XZ = 13 and XY = 5 calculate the values of YZ, PY, PX, and PZ.

(NCTEC)

38. (a) A wall is 7·0 m high. A ladder is placed with its foot 2·0 m from the foot of the wall. Calculate to the nearest 0·5 m the length of the ladder required to reach the top of the wall allowing 1·0 m extra for safety.

(b) A man 1·8 m tall stands on horizontal ground and a lamp 7·0 m high casts a shadow of the man 3·0 m long. What is the distance from the man's foot to the foot of the lamp?

(NCTEC)

39. Two gear wheels, of diameter 50 mm and 420 mm, are meshing together. If the larger gear wheel rotates through an angle of $2\pi/3$ radians, find the number of revolutions made by the smaller wheel.

(WJEC)

40. The bracket shown in Fig. R 4 had to be made from a 6 mm thick strip. Calculate the length of strip required at the mean thickness, that is the length of the centre line shown in the diagram.

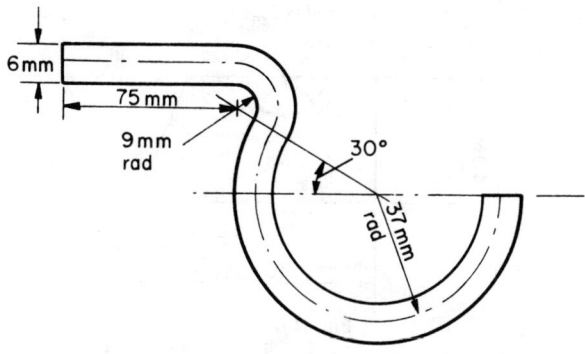

Fig. R 4

41. Two ball bearings of diameters 30 mm and 20 mm are dropped into a hollow cylinder standing on a horizontal table, as shown in Fig. R 5. The distance from the top of the upper bearing to the table is 40 mm. What is the internal diameter of the cylinder?

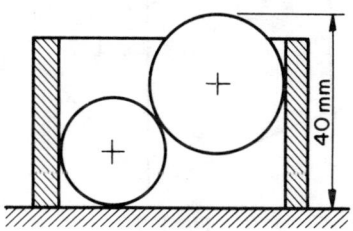

Fig. R 5

Section E: Areas and volumes

42. A water storage tank is in the form of an inverted cone surmounted by a cylinder, the diameter of the cylinder and the cone being 2 m. The vertical heights of the cylindrical portion and the conical portion of the tank are 3 m and 2·5 m respectively.

    Determine the volume of the tank.

    Give the answer correct to **two** decimal places.

    (UEI)

43. Figure R 6 shows the cross section of an extruded alloy bar, determine:
    (a) the cross sectional area, in mm².
    (b) the mass of a 1 m length of the bar if 1 m³ of the material has a mass of 1000 kg.

    (UEI)

271

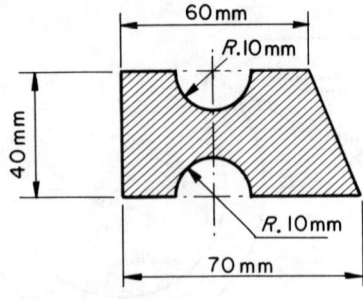

Fig. R 6

44. Figure R 7 shows a cube of 7 cm side with a hole 2 cm diameter passing through it. Determine (a) the total surface area and (b) the volume of material in the cube.

(EMEU)

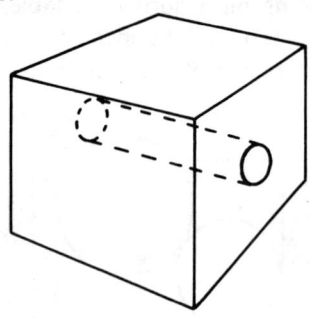

Fig. R 7

45. A block of wood 8 cm by 20 cm by 100 cm has two triangular prisms sawn off, as shown in Fig. R 8. AB is 8 cm and CD is 6 cm. Calculate:
    (i) the volume of the remaining wood
    (ii) its surface area.

(NCTEC)

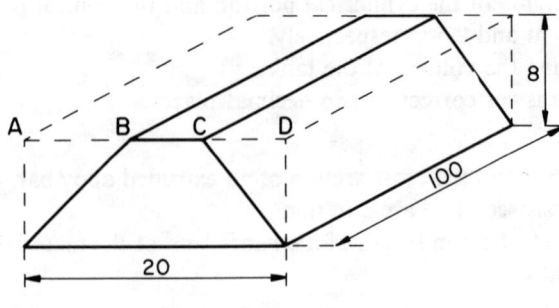

Fig. R 8

46. (a) Calculate the volume of a right circular cone of base diameter 16 m and slant height 10 m.
   (b) A steel rivet is in the shape of a cylinder surmounted by a hemi-spherical head. The diameter of the cylinder is 40 mm and that of the head 60 mm, the overall length of the rivet being 100 mm.
   Calculate the mass of 100 rivets given that 1 m³ of steel has a mass of 7850 kg.

(YHCFE)

47. (a) Figure R 9 shows a square inscribed in a circle. Find an expression for the area of the square in terms of the diameter D.
   (b) If the diameter of the circle is 14 cm calculate the area of the shaded portion. (Take π as $\frac{22}{7}$.)

(EMEU)

Fig. R 9

48. (a) The symmetrical cross section of a concrete footbridge is shown in Fig. R 10. The length of the bridge is 40 m. If 1 m³ of concrete weighs 5·4 kN and the bridge weighs 1485 kN determine the dimension x.
   (b) Calculate the surface area of the bridge.

(EMEU)

Fig. R 10

49. (a) An aluminium ingot has a volume of 990 mm³, the ingot is to be made into rod and then drawn into wire 0·75 mm diameter. Determine the length of wire produced.
   (b) A plumb bob is in the form of a cone of base diameter 5 cm and height 5 cm. Determine the mass of the plumb bob if it is made from material having a density of 8·85 g/cm³.

(UEI)

273

50. A hot water tank in the form of a cylinder has outside diameter 0·4 m and length 1·2 m. It is to be covered all over with an insulating material to a depth of 30 mm. Calculate the volume of insulating material required.

(ULCI)

51. An ingot 6 m long has a square cross section of 80 mm side. The ingot is to be made into ball bearings whose diameters are 14 mm. Determine the number of ball bearings produced.

(NCTEC)

Section F: Trigonometry

52. (a) In a triangle ABC, angle A = 90°, AB = 80 mm and BC = 250 mm. D is a point on AC such that angle ABD = 48°. Calculate
    (i) the length of AD correct to **two** decimal places
    (ii) the angle ACB.
    (b) In a triangle ABC, angle B = 42°, AB = 210 mm and AC = 380 mm. Calculate
    (i) the perpendicular distance from A to BC.
    (ii) the angle ACB.

(UEI)

53. Two cables, 20 m and 10 m long, are required for the temporary support of a concrete column as shown in Fig. R 11. Determine the distance between the anchor points A and B.

(EMEU)

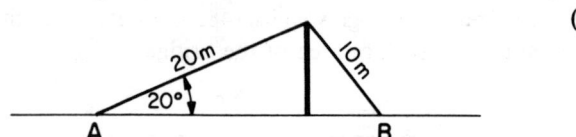

Fig. R 11

54. (a) Calculate the value of dimension A that should be obtained when checking the symmetrical dovetail shown in Fig. R 12.
    (b) If θ is an acute angle and sin $\theta = \frac{4}{5}$, find the values of cos θ and tan θ **without** using trigonometrical tables.

(ULCI)

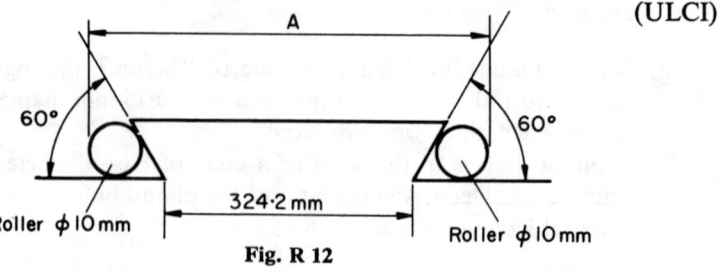

Fig. R 12

55. A triangle contains angles of 54° and 36°. Calculate the value of the third internal angle and the length of the longest side if the other two sides are 2·5 m and 6·539 m.

(NCTEC)

56. Assuming that the legs of a pair of dividers are 125 mm long and that they are set to scribe out a circle of 80 mm diameter:
    (i) Determine the included angle of the legs.
    (ii) If the included angle is now adjusted to 12° 38′ what diameter of circle could now be scribed out?

(NCTEC)

57. The diagonal of a rectangle 7 m by 10 m makes an angle X with the longer sides. Calculate (i) angle X, (ii) the length of the diagonal.

(NCTEC)

58. Figure R 13 shows a wire placed between the threads of a bolt. If the pitch of the thread (that is, the distance AB) is 3·0 mm, find the diameter of the wire, if the top of the wire is level with the tops of the threads.

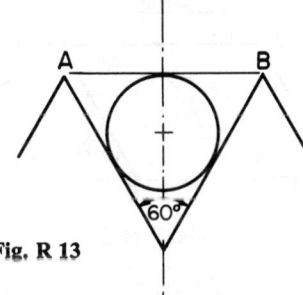

Fig. R 13

59. Three transistors are to be positioned on an aluminium base in the positions shown in Fig. R 14. If O is the centre of the circle, find the co-ordinates x and y of the top transistor.

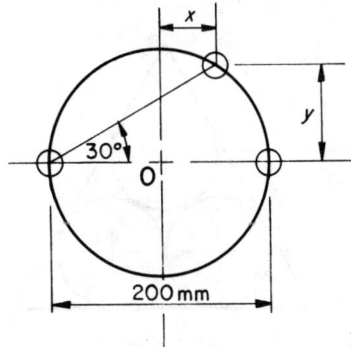

Fig. R 14

60. Calculate the taper α in Fig. R 15.

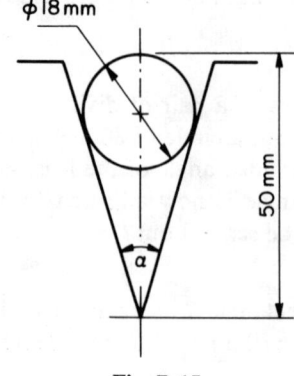

Fig. R 15

61. Find the cross-sectional area of the bar in Fig. R 16.

(ULCI)

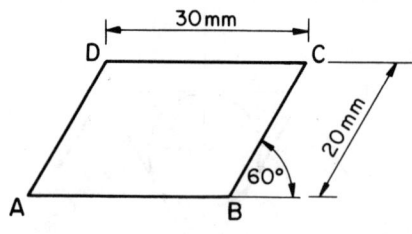

Fig. R 16

62. Three cylindrical plugs, each of diameter 40 mm, are placed in a equilateral triangular hole, each cylinder touching each other and being in contact with a side of the hole, as shown. Determine, by calculation, the length of the side of the triangular hole in mm, giving your answer to three significant figures (Fig. R 17).

(UEI)

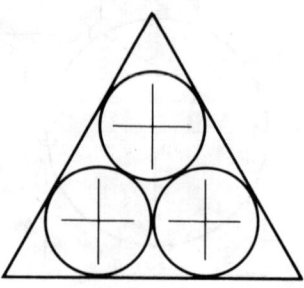

Fig. R 17

Section G: General

63. A wire-forming machine produces components as shown in Fig. R 18. Calculate how many components can be produced from 50 m of wire if 5% of the wire is wasted in cutting off the components.

   (*A solution based upon drawing is not acceptable.*)
   (ULCI)

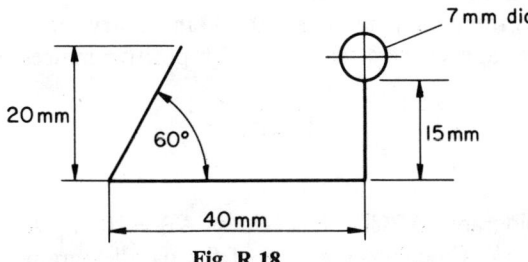

Fig. R 18

64. A machine produces blanks of hexagonal cross section, 24 mm across flats by 10 mm thick, with a hole of diameter 12 mm through the centre.

   Using the given extract from British Standard tables, determine the mass of 1000 blanks.

Hexagon, across flats (mm)	22	24	26
Mass (kg per m)	3·29	3·62	4·60
Rounds diameter (mm)	10	12	14
Mass (kg per m)	0·62	0·89	1·21

   (ULCI)

65. (a) Simplify the expression

$$\sqrt{\left(\frac{25x^6}{y^4}\right)}$$

   and determine its value when $x = 2$ and $y = 5$.

   (b) Express 27 in binary form.

   (c) Express the binary number 110110 in denary form.

   (UEI)

66. (a) The interior angles of a triangle are $(80-x)°$, $(4x-5)°$, and $(2x-15)°$. Determine the magnitude of the three angles.

   (b) A right circular cone has a radius of 10 cm and height 8 cm. Calculate
   (i) its volume
   (ii) the radius of a cylindrical container of length 4 cm which has the same volume.

   (EMEU)

277

67. The braking surface of a brake lining is in the form of an arc of a circle of radius 12 cm and the angle subtended by the arc is 120°.

Calculate the length of the braking surface, in centimetres, giving the answer correct to **one** place of decimals.

(YHCFE)

68. (a) Express 41 in binary form.

(b) Express the binary number 1100100 in denary form.

(c) Simplify and express the result with positive indices only.

$$\left(\frac{a^{-\frac{1}{4}}b}{c^{-\frac{1}{4}}}\right)^{-2}$$

(UEI)

69. A parallelogram ABCD has sides AB = 20 m, AD = 25 m, and ∠ A = 13° 54'. Calculate the area of the parallelogram.

(YHCFE)

70. Figure R 19 shows the section of an open gutter 30 m long. Determine the area of sheet metal required to make this gutter.

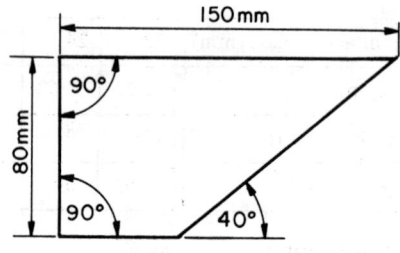

Fig. R 19

71. (a) The stress in a particular tie bar is 8000 N/mm². Due to corrosion its cross-sectional area is reduced causing the stress to increase to 11 000 N/mm².

Calculate the percentage increase in stress.

(b) Determine the area of an equilateral triangle whose sides are of length 250 mm.

(c) A rectangle has a width of 120 mm and a diagonal of 260 mm. Calculate the length of the rectangle.

(UEI)

72. A square is to be machined on the end of a shaft as shown in Fig. R 20. Calculate the maximum possible size of square.

(ULCI)

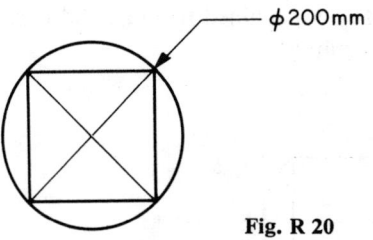

$\phi 200\text{mm}$

Fig. R 20

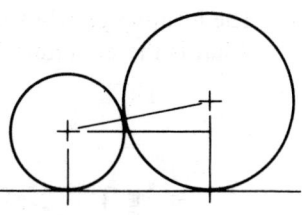

Fig. R 21

73. Figure R 21 shows two ball bearings in contact with each other on a horizontal table. The diameters of the balls are 8 mm and 12 mm. What is the distance between their points of contact with the table?

74. A cylindrical bar, diameter 100 mm, rests in a 90° Vee block as shown in Fig. R 22. From the dimensions shown find the distance from the bottom of the Vee-notch to the bottom of the block.

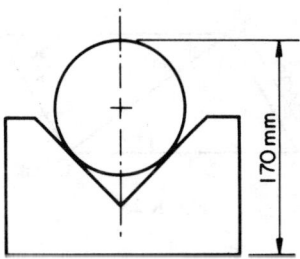

170 mm

Fig. R 22

75. Two gear wheels have their centres on a line inclined at 60° to the horizontal axis, as shown in Fig. R 23. Find the horizontal and vertical distances x and y between the centres.

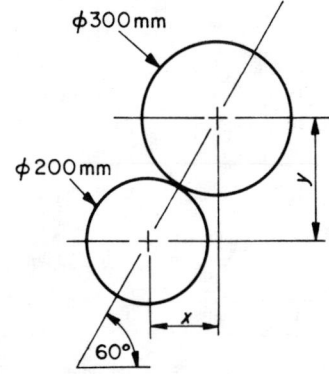

$\phi 300\text{mm}$

$\phi 200\text{mm}$

y

x

60°

Fig. R 23

279

76. The locating pin shown in Fig. R 24 is to be machined from a solid bar. What is the minimum length L of bar required?

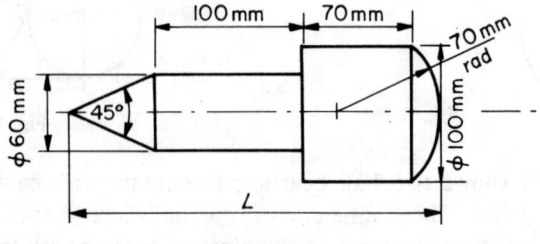

Fig. R 24

77. Find the distance AC in the template ABCD, which is in the form of a parallelogram, shown in Fig. R 25.

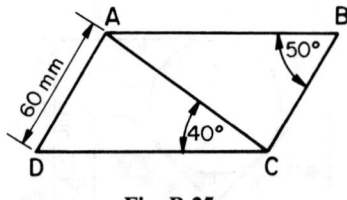

Fig. R 25

Answers to exercises

Exercise 1.1
1. 9 **2.** 6 **3.** 6 **4.** 10 **5.** 35 **6.** 12 **7.** 210 **8.** 5 **9.** 22 **10.** 40

Exercise 1.3
1. (a) 2 (b) 2, 5 (c) 3 (d) 2, 3, 5 (e) 3 **2.** 3, 5, 11, 13, 17
3. (a) $2 \times 2 \times 2 \times 2 \times 3$ (b) $2 \times 2 \times 19$ (c) $2 \times 2 \times 2 \times 2 \times 2 \times 2 \times 2$
    (d) $2 \times 2 \times 2 \times 5 \times 23$ (e) $2 \times 2 \times 3 \times 3 \times 3$ (f) $2 \times 3 \times 5 \times 7$ (g) $3 \times 5 \times 5 \times 7$
4. (a) $2 \times 2 \times 2 \times 2 \times 2 \times 2 \times 2 \times 3 \times 7 \times 7$ (b) $3 \times 7 \times 11 \times 13 \times 37$
    (c) $3 \times 7 \times 13 \times 13 \times 37$

Exercise 1.4
1. (a) 14 (b) 30 (c) 12 (d) 28 (e) 63 (f) 252 (g) 42 (h) 105
2. (a) 6 (b) 6 (c) 2 (d) 2 (e) 3 (f) 2 (g) 2 (h) 5
    (b) (a) 4 (b) 9 (c) 20 (d) 3 (e) 2 (f) 28 (g) 5 (h) 2
3. (a) 9 (b) 6 (c) 4 (d) 12 (e) 14 (f) 8 (g) 10 (h) 21

Exercise 1.5
1. 30 **2.** 12 **3.** 6 **4.** 10 **5.** 8 **6.** 9 **7.** 30 **8.** 105 **9.** 4 **10.** 5

Exercise 1.6
1. 180 **2.** 225 **3.** 162 **4.** 1470 **5.** 22 050 **6.** 330 **7.** 252 **8.** 1260
9. 1260 **10.** 2310

Exercise 1.7
2. (a) 1 (b) 3 (c) 5 (d) 7 (e) 5 (f) 12 (g) 50 (h) 30 (i) 21 (j) 2
3. (a) $\frac{1}{3}$ (b) $\frac{2}{5}$ (c) $\frac{2}{5}$ (d) $\frac{4}{3}$ (e) 5
4. (a) $\frac{11}{4}$ (b) $\frac{27}{4}$ (c) $\frac{89}{21}$ (d) $\frac{47}{10}$ (e) $\frac{11}{6}$ (f) $\frac{107}{10}$ (g) $\frac{131}{11}$ (h) $\frac{3}{1}$ (i) $\frac{5}{1}$ (j) $\frac{10}{1}$
5. (a) $1\frac{5}{9}$ (b) $2\frac{1}{10}$ (c) $3\frac{4}{7}$ (d) $1\frac{5}{6}$ (e) $11\frac{1}{9}$ (f) $1\frac{4}{5}$ (g) $1\frac{6}{13}$ (h) 2 (i) 2
    (j) 7 (k) 121

Exercise 1.8
1. (a) $\frac{2}{5}, \frac{1}{6}, \frac{2}{15}$ (b) $\frac{5}{12}, \frac{3}{8}, \frac{1}{3}$ (c) $\frac{2}{3}, \frac{7}{12}, \frac{1}{4}$ (d) $\frac{3}{4}, \frac{9}{14}, \frac{4}{7}$ (e) $\frac{4}{5}, \frac{3}{10}, \frac{7}{25}$
2. (a) $\frac{29}{35}$ (b) $1\frac{5}{12}$ (c) $1\frac{1}{5}$ (d) $\frac{35}{66}$ (e) $6\frac{17}{36}$ (f) $3\frac{63}{80}$ (g) $5\frac{43}{105}$ (h) $9\frac{11}{20}$ (i) $8\frac{1}{36}$
3. (a) $\frac{1}{16}$ (b) $\frac{13}{25}$ (c) $\frac{1}{18}$ (d) $1\frac{5}{12}$ (e) $3\frac{19}{60}$ (f) $2\frac{5}{44}$ (g) $1\frac{9}{20}$ (h) $3\frac{23}{48}$ (i) $\frac{7}{8}$
4. (a) $3\frac{7}{16}$ (b) $\frac{43}{60}$ (c) $3\frac{5}{12}$ (d) $1\frac{7}{8}$ (e) $2\frac{37}{50}$

Exercise 1.9
1. $\frac{1}{8}$ **2.** $\frac{2}{3}$ **3.** $\frac{1}{8}$ **4.** 3 **5.** $7\frac{1}{2}$ **6.** 25 **7.** $\frac{6}{7}$ **8.** $\frac{4}{5}$ **9.** $1\frac{1}{2}$ **10.** $2\frac{2}{5}$ **11.** $1\frac{1}{3}$
12. 4 **13.** 4 **14.** $4\frac{1}{2}$ **15.** $\frac{3}{40}$ **16.** $2\frac{47}{64}$ **17.** $16\frac{1}{3}$

Exercise 1.10
1. $\frac{1}{2}$ **2.** $\frac{19}{20}$ **3.** $\frac{7}{18}$ **4.** $1\frac{8}{13}$ **5.** $10\frac{5}{12}$ **6.** $4\frac{6}{25}$ **7.** $1\frac{1}{2}$

Exercise 1.12
1. 147·6 **2.** 13 600 **3.** 0·71 **4.** 712·3 **5.** 1471·2 **6.** 700 **7.** 2·14
8. 10 101·01 **9.** 160 **10.** 71·72 **11.** 40 **12.** 0·4 **13.** 0·027 12 **14.** 6·12
15. 0·716 **16.** 0·0061 **17.** 0·000 602 **18.** 0·014 613 **19.** 0·1617 **20.** 0·0716

Exercise 1.13
1. $\frac{3}{5}$ **2.** $\frac{4}{5}$ **3.** $\frac{9}{100}$ **4.** $\frac{2}{25}$ **5.** $\frac{9}{20}$ **6.** $\frac{18}{25}$ **7.** $\frac{33}{100}$ **8.** $\frac{5}{8}$ **9.** $\frac{33}{40}$ **10.** $6\frac{11}{25}$
11. $\frac{1}{100}$ **12.** $\frac{1}{10000}$ **13.** 0·3 **14.** 0·7 **15.** 0·42 **16.** 0·36 **17.** 0·375
18. 0·24 **19.** 0·35 **20.** 0·3125 **21.** 0·6875 **22.** 0·9375 **23.** 0·218 75
24. 0·0875

Exercise 1.14
1. 7·2 **2.** 16·76 **3.** 21·70 **4.** 0·060 **5.** 14 **6.** 17·47 **7.** 108·465 **8.** 0·0
9. 0·1 **10.** 10·06

Exercise 1.15
1. 71700, 46·6, 392, 1·07, 0·00625 **2.** 30, 1·3, 1·1, 0·17, 9·1, 0·071

Exercise 1.16
1. 426·676 **2.** 144·6692 **3.** 4130·852 **4.** 0·790 32 **5.** 0·212 110 1
6. 13·54 **7.** 0·0172 **8.** 0·5805 **9.** 9·283 **10.** 0·9396 **11.** 3·52 **12.** 204·82
13. 7·08 **14.** 41·78 **15.** 8·72 **16.** 0·10 **17.** 4·46 **18.** 15·84

Exercise 2.1
1. (a) 3 : 2 (b) 4 : 5 (c) 2 : 3 (d) 2 : 3 : 4 (e) 4 : 6 : 5 (f) 6 : 5 (g) 5 : 3
(h) 25 : 4 **2.** 60 g **3.** 180 W **4.** $\frac{1}{2}$ m, 2 m, 1$\frac{1}{2}$ m, **5.** 9 cm, 12 cm, 15 cm
6. 6 : 1 : 19 **7.** 7 : 8 **8.** 9 kg, 36 kg, 54 kg **9.** 8 kg, 28 kg **10.** (a) 1 : 3 : 4
(b) 2 : 3 : 5

Exercise 2.2
1. 900 N/m² **2.** 11 gallons **3.** 600 J **4.** £93·75 **5.** 300 miles **6.** 15 cm
7. 54 mph **8.** 32 rev/m **9.** 13·5 h **10.** 3·8 A **11.** 20 men

Exercise 2.3
1. 10% **2.** 12% **3.** 62·5% **4.** 76% **5.** 33$\frac{1}{3}$% **6.** 6$\frac{1}{4}$% **7.** 37$\frac{1}{2}$%
8. 41$\frac{2}{3}$% **9.** 31% **10.** 52% **11.** 67% **12.** 5% **13.** $\frac{1}{2}$% **14.** 11$\frac{1}{2}$%
15. 0·3, $\frac{3}{10}$ **16.** 0·45, $\frac{9}{20}$ **17.** 0·67, $\frac{2}{3}$ **18.** 0·9, $\frac{9}{10}$ **19.** 0·8, $\frac{4}{5}$ **20.** 0·06, $\frac{3}{50}$
21. 0·005, $\frac{1}{200}$ **22.** 0·000 125, $\frac{1}{8000}$

Exercise 2.4
1. 25% **2.** 105 **3.** £138 **4.** 12·5% **5.** (a) 79 A, 81 A (b) 14·4 A, 15·6 A
6. 28·2 kg, 3·2 kg, 0·6 kg **7.** 4·2% **8.** (a) 6·25% (b) 203 V
9. 1·11 ohm, $\frac{111}{1441}$, 7·7%

Exercise 2.5
1. 10^5 **2.** 4^8 **3.** 3^{10} **4.** 9^6 **5.** 4^2 **6.** 9^2 **7.** 4^2 **8.** 8×6 **9.** 7×4^3
10. $(\frac{1}{2})^2$ **11.** $(\frac{1}{10})^3$ **12.** $\frac{1}{8}$ **13.** 2^4 **14.** 10^3 **15.** 9 **16.** 1 **17.** 1 **18.** 1
19. 4 **20.** 6 **21.** 1 **22.** 1 **23.** 1 **24.** 2^6 **25.** 3^8 **26.** 10^{15} **27.** 9^{21} **28.** 6^{14}
29. 6^{26} **30.** 2^{23} **31.** 9 **32.** 8 **33.** $\frac{1}{4}$ **34.** $\frac{3}{8}$ **35.** $\frac{5}{7}$ **36.** $\frac{5}{3}$ **37.** $\frac{3}{2}$ **38.** 3
39. $\frac{1}{4}$ **40.** $\frac{3}{4}$ **41.** $\frac{2}{5}$ **42.** $\frac{5}{4}$

Exercise 2.6
1. $3·611 \times 10^2$, $4·216 \times 10^3$, $3·362 \times 10$ **2.** $3·67 \times 10^{-3}$, $2·17 \times 10^{-1}$, $4·17 \times 10^{-2}$
3. (a) $2·25 \times 10^5$ (b) 6×10^5 **4.** (a) 361·2 (b) 4316 (c) 26 000 (d) 0·0581
(e) 0·6217 (f) 0·0 000 822

Exercise 2.7
1. $5 \cdot 295 \times 10^3$ **2.** $1 \cdot 125 \times 10^{-2}$ **3.** $12 \cdot 07 \times 10^{-4}$ **4.** $1 \cdot 91 \times 10^2$ **5.** $5 \cdot 945 \times 10^2$
6. $37 \cdot 63 \times 10^2$ **7.** $16 \cdot 85 \times 10^{-3}$ **8.** $60 \cdot 97 \times 10^{-3}$ **9.** $68 \cdot 92 \times 10^{-5}$
10. $5 \cdot 0005 \times 10^2$

Exercise 2.8
1. (a) 55 (b) 5 (c) 19 (d) 33 (e) 16 (f) 26 (g) 37 (h) 27
2. (a) 1011 (b) 11000 (c) 1010 (d) 11101 (e) 10011 (f) 1000000
(g) 100000 (h) 101101 **3.** (a) 111 (b) 1000 (c) 11001 (d) 100010
(e) 1001101 (f) 100000

Exercise 3.1
1. $1 \cdot 510$, $28 \cdot 91$, $158 \cdot 1$, $0 \cdot 5805$, $0 \cdot 2300$
2. $1 \cdot 242$, $6 \cdot 993 \times 10^2$, $3 \cdot 36 \times 10^{-2}$, $8 \cdot 593 \times 10^{-3}$ **3.** $0 \cdot 3571$, $0 \cdot 002930$, $1 \cdot 016$, $157 \cdot 8$
4. (a) $0 \cdot 06656$ (b) $0 \cdot 1322$ (c) $16 \cdot 10$
5. (a) $6 \cdot 632$ (b) $0 \cdot 1008$ (c) $6 \cdot 300$ (d) $0 \cdot 5587$ (e) $2 \cdot 083$

Exercise 3.2
1. 1, 3, 0, 1, 2 **2.** $0 \cdot 3522$, $0 \cdot 5652$, $0 \cdot 9262$, $0 \cdot 9991$, $0 \cdot 7344$
3. $1 \cdot 2810$, $2 \cdot 8495$, $0 \cdot 6473$, $3 \cdot 6654$, $1 \cdot 9367$

Exercise 3.3
1. 141 100, $5 \cdot 014$, $10 \cdot 28$, $812 \cdot 8$, 19 950
2. 1348, $277 \cdot 7$, $10 \cdot 22$, $1 \cdot 996$, $123 \cdot 3$, $1 \cdot 023$

Exercise 3.4
1. (a) $7 \cdot 145 \times 10$ (b) $1 \cdot 262 \times 10^3$ (c) $1 \cdot 277 \times 10^5$ (d) $1 \cdot 229 \times 10^5$
2. (a) $2 \cdot 70$ (b) $1 \cdot 03$ (c) 945 (d) 1830
3. (a) 579 300 (b) $46 \cdot 43$ (c) $1 \cdot 160$ (d) $1 \cdot 310$
4. (a) $7 \cdot 78$ (b) $2 \cdot 37$ (c) 134 (d) $6 \cdot 28$ (e) $22 \cdot 5$

Exercise 3.5
1. (a) $4 \cdot 815$ (b) $2 \cdot 701 \times 10$ (c) $2 \cdot 41$ (d) $4 \cdot 147 \times 10^{-3}$ (e) $5 \cdot 683 \times 10^{-5}$
2. (a) $38 \cdot 74$ (b) $0 \cdot 001 261$ (c) $5 \cdot 196$ (d) $0 \cdot 5501$ (e) $0 \cdot 3426$
3. (a) $1 \cdot 766$ (b) $1 \cdot 301$ (c) $73 \cdot 16$ **4.** (a) $1 \cdot 45$ (b) $70 \cdot 0$ (c) $3 \cdot 52$ (d) $44 \cdot 6$

Exercise 3.6
1. $0 \cdot 3571$ **2.** $0 \cdot 002929$ **3.** $1 \cdot 017$ **4.** $157 \cdot 8$

Exercise 3.7
1. $2 \cdot 235$, $3 \cdot 472 \times 10^2$, $8 \cdot 579 \times 10^3$, $2 \cdot 295 \times 10^4$, $4 \cdot 163 \times 10^6$
2. $4 \cdot 497 \times 10$, $7 \cdot 984 \times 10^2$, $3 \cdot 051 \times 10^3$, $1 \cdot 418 \times 10^5$
3. $1 \cdot 15$, $3 \cdot 59$, $12 \cdot 7$, 250 **4.** (a) $4 \cdot 266$ (b) $2 \cdot 535$ (c) $40 \cdot 20$ (d) $1 \cdot 268$ (e) $47 \cdot 25$

Exercise 3.8
1. (a) $0 \cdot 1945$ (b) $0 \cdot 3381$ (c) $0 \cdot 2798$ (d) $0 \cdot 05295$ (e) $0 \cdot 6921$ (f) $0 \cdot 4978$
2. (a) $0 \cdot 02028$ (b) $0 \cdot 07023$ (c) $0 \cdot 000458$ (d) $0 \cdot 8414$

Exercise 4.1
1. $7x$ **2.** $2p$ **3.** Not possible **4.** Not possible **5.** $14t$ **6.** $2c$ **7.** $10e$
8. $12y + 3a$ **9.** $8t + 3m$ **10.** $2a + 8b$ **11.** $12p + 7q$ **12.** z **13.** 0

Exercise 4.2
1. $8ab$ **2.** $3vxz$ **3.** Not possible **4.** Not possible **5.** Not possible **6.** $12xy$
7. ut **8.** $4rtv$ **9.** $12axy$ **10.** $2xy+11mn$ **11.** $3ab+12uv$ **12.** $9xy+6af$
13. $14ft$

Exercise 4.3
1. $5p+5q$ **2.** $7u+7v-7n$ **3.** $lt+mt+nt$ **4.** $-5r-5s+5t$ **5.** $2ay-2az$
6. $2ar+2br-6cr$ **7.** $mnpq-mnrs$ **8.** $4xy+4xz$ **9.** $-2xy-2yz$
10. $5abz-5bcz$

Exercise 4.4
1. $35x$ **2.** $2t$ **3.** $9yz$ **4.** 0 **5.** $3t$ **6.** $7z$ **7.** $144uvw$ **8.** $4xz$ **9.** uvw
10. $10cdf.$

Exercise 4.5
1. $19m$ **2.** $4p$ **3.** $11p$ **4.** $14f$ **5.** $12f$ **6.** $16t$ **7.** $55a$ **8.** $8rs$ **9.** $11mn$
10. 0

Exercise 4.6
1. 9 **2.** 9 **3.** 17 **4.** 9 **5.** 120 **6.** 2 **7.** 8 **8.** $\frac{13}{12}$ **9.** $\frac{25}{12}$

Exercise 4.7
1. -3 **2.** -7 **3.** $-2x$ **4.** $-9m$ **5.** $-30t$ **6.** 24 **7.** -20 **8.** -56
9. -20 **10.** $12y$ **11.** $40s$ **12.** -3 **13.** -3 **14.** 7 **15.** $-2m$ **16.** $-7r$
17. 2 **18.** -2 **19.** -12 **20.** $9z$ **21.** $-12p$ **22.** a

Exercise 4.8
1. (a) x^4 (b) 3^4 (c) $(2y)^4$ (d) x^{10} **2.** (a) $a\times a$ (b) $x^2\times x^2\times x^2$ (c) $2x\times 2x$
3. (a) p^9 (b) $6m^9$ (c) $-18t^6$ (d) $20y^5$ (e) $30x^{11}$ (f) $60x^6$ (g) $-48f^9$
4. (a) 1 (b) $-9f^2$ (c) $\frac{4}{3}e^4$ (d) $-\frac{3}{5}q$ **5.** (a) x^6 (b) $-32v^{20}$ (c) $49w^6$ (d) 1
(e) 81

Exercise 4.9
1. $4x^4$ **2.** $10x^{-\frac{1}{3}}$ **3.** $14/t$ **4.** t^{-10} **5.** $1/m$ **6.** p^{-8} **7.** $\frac{3}{2}p^7$ **8.** $\frac{3}{8}p$ **9.** $2x^4$
10. $4x^4$ **11.** $3x^4$ **12.** $27x^4$ **13.** $\frac{1}{16}$ **14.** $\frac{27}{64}$ **15.** $\frac{1}{32}$

Exercise 4.10
1. $3a+6b+9c$ **2.** $-3x+12$ **3.** $-2a+b+c$ **4.** $s+2t$ **5.** $-11a+2b+3c$
6. $51x+150$ **7.** x^4-x^2 **8.** $36x^3-12x^2+4x$ **9.** $3a-b+2c$

Exercise 4.11
1. x^2+3x+2 **2.** y^2+y-12 **3.** $2z^2-10z+12$ **4.** $-3x^2+8x+35$
5. $x^2-xy-2y^2$ **6.** $a^2+4ab-5b^2$ **7.** $y^2+16y+64$ **8.** $x^2+14x+49$
9. $y^2+4uv+4u^2$ **10.** x^2-y^2 **11.** v^2-16u^2 **12.** $54t^2-39st-28s^2$

Exercise 4.12
1. (a) $a(x+3y+4z)$ (b) $3t^2(1-3t)$ (c) $mn(x+y)$ (d) $t(f-g)$ (e) $a^2(x-y)$
(f) $r(a-b-c)$ (g) $5(1-p)$ (h) $4t^2(t^2+8t+2)$
2. (a) $-(x-y)$ (b) $-(3z-4y+7x)$ (c) $-(-3x+4)$
3. (a) $(4-z)(x+y)$ (b) $(x+1)(x+4)$ (c) $(4m+n)(l-4)$ (d) $(x-6)(4l+3)$
(e) $(m+3)(4m-3)$ (f) $(x-y)(4t+1)$
4. (a) $(b-c)(a+e)$ (b) $(u-v)(t-s)$ (c) $(x+5y)(5a-5b)$ (d) $(x-6)(x+3)$
(e) $(t-7)(t-4)$ (f) $(x-a)(x-1)$

284

Exercise 5.1
1. 4   **2.** $2\frac{1}{4}$ **3.** 8   **4.** 5   **5.** 7   **6.** $-\frac{10}{3}$ **7.** $\frac{1}{2}$ **8.** $-8\frac{1}{2}$

Exercise 5.2
1. -1 **2.** $-9\frac{1}{2}$ **3.** 5   **4.** 29   **5.** $\frac{114}{7}$ **6.** 32   **7.** 1   **8.** 24

Exercise 5.3
1. £530, £1060   **2.** $7m, 4m$ **3.** 23, 24, 25   **4.** £270   **5.** 17 per h   **6.** 8 A, 2 A
7. 1000 h   **8.** £9   **9.** $\frac{2}{M} \times 10^3$ **10.** 60, 15   **11.** 9 cm

Exercise 5.4
1. 3, 1   **2.** 6, 1   **3.** $\frac{1}{7}, \frac{11}{7}$ **4.** 2, 2   **5.** $-\frac{5}{2}, -1$ **6.** 5, -1 **7.** $-4, 5$
8. $-4, 2$

Exercise 5.5
1. 4·4   **2.** 360   **3.** 50·08, 0·08   **4.** 0·13, 1·87   **5.** 10·0   **6.** 0·8   **7.** -40
8. 36 240   **9.** 61·08   **10.** 240, no change

Exercise 5.6
1. $b = \dfrac{(P-2a)}{2}$ **2.** $a = A/h - b$ **3.** (a) $z = \dfrac{(b-a)}{2}$ (b) $z = c/2ab$ (c) $\dfrac{3-x-y}{2b}$

(d) $z = \dfrac{6}{(a+b)}$ (e) $z = \dfrac{r}{(x+y)}$ **4.** $r = 1 - a/s$ **5.** $a = \dfrac{(v-u)}{t}$

6. $l = T^2 g/4\pi^2$ **7.** $v = \sqrt{(2E/m)}$ **8.** $l = YAx/F$ **9.** $r = \sqrt[3]{(3V/4\pi)}$

10. $\sqrt{(a^2 - v^2/w^2)}$ **11.** $r = \dfrac{VR}{(E-V)}$ **12.** $\alpha = \dfrac{(L-l)}{lt}$ **13.** $r = \sqrt[4]{\left(\dfrac{\pi R^4 - M}{\pi}\right)}$

Exercise 6.1
1. (a) 45 m   (b) $3\frac{1}{2}$ s   **2.** (a) $13\frac{1}{2}$ V   (b) 6 A   **3.** (a) 840 J   (b) $68\frac{1}{3}$
4. (a) 175 N   (b) 25 mm   **5.** 40   **6.** 8

Exercise 6.4
Gradients: **1.** 10   **2.** 3   **3.** 120   **4.** 5   **5.** 40   **6.** 8

Exercise 8.1
1. (a) $\alpha = 240°$, (b) $\alpha = 65°$, $\beta = 35°$, $\gamma = 105°$ (c) 58°, (d) 100°   **2.** (a) (i) 41°
(ii) 52° 39′ (iii) 10° 54′ (b) (i) 102° (ii) 99° 48′ (iii) 32° 11′   **3.** 331°   **4.** 60°
5. 18°   **6.** $62\frac{1}{2}°$ **7.** 65° or 115°   **9.** (a) 36°   (b) 225°   **10.** 122°, 58°, 122°

Exercise 8.2
1. (a) 66°, 24°   (b) 50°, 93°   (c) 34°, 22°   (d) 50°   **2.** 40° 48′   **3.** 61°, scalene
4. 30°, 30°   **5.** 66°, 56°   **6.** 15°   **7.** (a) $\alpha = 60°$, $\beta = 50°$, $\theta = 70°$ (b) $\alpha = 60°$
8. 48°, 55°, 77°   **9.** 33°,   **10.** 25°

Exercise 8.3
1. (a) 17·2   (b) 5·2   **2.** 141 mm   **3.** 95·7 mm   **4.** 11·3 mm   **5.** 12·2m
6. 30·3 m   **7.** 199 mm

Exercise 8.4
1. 161 mm   **2.** 20·0 m

Exercise 8.6
2. 10·6   **4.** 17·1 m   **5.** 11·7 mm   **6.** 1·4 m, 8·4 m   **7.** 4·5, 7·5   **8.** 3·43 mm
9. 8·88 mm   **10.** 11·6 m   **11.** 728 mm

Exercise 9.1
1. (a) 628 mm   (b) 2·9 m   (c) 5·3 m   (d) 10·2 m   **2.** 6·17 m   **3.** 1·43 m
4. 9·43 m   **5.** (a) 446   (b) 246 mm   **6.** 0·245   **7.** 19 mm   **8.** 3·93 mm

Exercise 9.2
1. 130°, 30°   **2.** 71°, 71°, 38°   **3.** 0·24 m   **4.** 55°   **5.** 80° 60°, 40°   **6.** 55°

Exercise 9.3
1. 0·92 m   **2.** 36·2 mm   **3.** 120 mm   **4.** 87 mm   **5.** 104 mm   **6.** 4·26 mm
7. 6·9 m   **8.** 451 mm

Exercise 9.4
1. (a) 180°   (b) 360°   (c) 90°   (d) 270°   (e) 60°   (f) 45°   (g) 120°   (h) 135°
(i) 150°   (j) 300°   (k) 540°   (l) 420°   **2.** (a) $\pi/6$ (b) $\pi/20$ (c) $\pi/18$
(d) $5\pi/4$ (e) $11\pi/6$ (f) $7\pi/3$ **3.** (a) 28° 39′   (b) 68° 45′   (c) 171° 54′
(d) 9° 45′   (e) 320° 52′   **4.** (a) 0·2094   (b) 0·3142   (c) 0·2443   (d) 1·8500
(e) 5·6898   **5.** (a) 0·4712   (b) 0·7374   (c) 1·4667   (d) 3·2434   (e) 5·2788
6. (a) 22° 55′   (b) 6° 43′   (c) 69° 30′   (d) 57° 46′   **7.** 22·0 mm, 166 mm,
8. 0·89 m   **9.** 135° 14′   **10.** 680 mm   **11.** 3·77 m   **12.** 183° 20′   **13.** 180

Exercise 9.5
1. (a) $\alpha = 95°$ (b) $\alpha = 105°$, $\beta = 45°$ (2a) $\alpha = 80°$, $\beta = 60°$ (b) $\alpha = 68°$,
$\beta = 138°$ **3.** 500 mm

Exercise 10.1
1. (a) 225 mm², 60 mm   (b) 0·01 m², 0·580 m
2. (a) 900 mm², 200 mm   (b) 1300 mm², 380 mm   (c) 1136 mm², 404 mm
3. (a) 0·164 m²   (b) 31 800 mm²   (c) 1200 mm²   (d) 4330 mm²
4. (a) 600 mm²   (b) 4·34 mm²   **5.** (a) 12 000 mm²   (b) 460 mm
6. (a) 69·3 mm   (b) 4·2, 12·7   **7.** 6·7 mm   **8.** 2·6 m², 2·9 m²   **9.** 2 : 3

Exercise 10.2
1. (a) $3·14 \times 10^4$ mm²   (b) 0·67 m²   (c) 2·3 m²   (d) 8·24 m²   **2.** 1·80 m²
3. 0·13 m²   **4.** 40·6 mm, 127 mm   **5.** 0·118 m², 14 000 mm, 13·6 m², 15·1 m,
4150 mm², 330 mm, 43 900 mm², 1110 mm, 4750 mm², 417 mm, 364 mm²,
156 mm, 0·492 m², 3·4 m   **6.** 160 m²   **7.** 4·20 m²   **8.** 149 500 mm²

Exercise 10.3
1. 64 000 mm³   **2.** (a) $3·15 \times 10^4$ mm³   (b) $2·6 \times 10^{-2}$ mm³   **3.** 3·4 m³
4. 1·75 m²   **5.** 0·026 m³   **6.** 0·0012 m³   **7.** $9·6 \times 10^5$ mm³   **8.** 0·000 37 m³

Exercise 10.4
1. (a) 1 130 000 mm³, 37 700 mm²   (b) 0·088 m³, 1·305 m²   (c) 0·0965 m³
0·78 m²   **2.** 0·20 m²   **3.** 320 mm, 0·10 m²   **4.** 20 mm   **5.** $1·8 \times 10^5$ mm³
6. 47·5 m²   **7.** 0·0025 m³   **8.** 0·082 m³   **9.** 216 mm

Exercise 10.5
1. (a) 4190 mm³ (b) 268 000 mm³ (c) 44·5 m³ **2.** 2·5 m **3.** 0·23 m³, 2·15 m²
4. 2·94 m³ **5.** 3·6 × 10⁴ mm³ **6.** 0·64 mm

Exercise 10.6
1. (a) 64 000 mm³ (b) 3460 mm³ **2.** (a) 16 500 mm³ (b) 314 mm³
(c) 960 000 mm³ (d) 226 mm³ **3.** 2260 mm³ **4.** 8·7 × 10⁴ mm³
5. 2950 mm³ **6.** 56·5 mm

Exercise 11.1
1. 0·766 **2.** 0·342 **3.** 0·727

Exercise 11.2
1. (a) 0·1736 (b) 0·3912 (c) 0·2954 (d) 0·3395 (e) 0·8895 (f) 0·9972
     0·9848   0·9303   0·9554   0·9406   0·4568   0·0758
     0·1763   0·4252   0·3093   0·3610   1·9472   13·15
2. (a) 73° 44′ (b) 0° 32′ (c) 44° 28′ (d) 83° 43′
3. (a) 63° 2′ (b) 44° 51′ (c) 20° 56′ (d) 88° 48′
4. (a) 24° 41′ (b) 54° (c) 65° 27′ (d) 81° 4′
5. (a) 5·8581 (b) 0·1997 (c) 0·803
6. 2·0323

Exercise 11.3

1. (a) $3\sqrt{3}$ (b) 3 (c) $\dfrac{5\sqrt{2}}{2}$ (d) $\dfrac{\sqrt{3}}{2}$

Exercise 11.4
1. (a) 39° 10′ (b) 44° 32′ (c) 56° 45′ **2.** 43° 47′, 46° 13′
3. (a) 37·8, 29·0 (b) 5·33, 5·53 (c) 1·66, 3·18 **4.** 3·49 m **5.** 17 mm, 14 mm
6. 65 mm **7.** 69° 12′, 69° 12′, 41° 36′ **8.** 41 mm, 69 mm **9.** 132 mm
10. 128 mm **11.** 6·2 mm **12.** 16·4 mm **13.** 44° 25′ **14.** 41 m **15.** 1·5 km
16. 1° 54′ **17.** 10·1 m **18.** 1·29 m, 76° 30′ **19.** 13 mm **20.** 1·84 m
21 10 m, 36° 52′ **22** 9° 32′ **23** 5·36 mm, 8·22 mm **24.** 27·61 mm, 126·6 mm

Revision Exercise
Section A
1. (a) (i) $\frac{8}{15}$ cm, (ii) 1·35 N (b) 61·3 rev/min
2. (a) 150 ohm, 225 ohm, 525 ohm (b) 346·5 ohm, 313·5 ohm (c) 100 s
3. (a) 72, 46·8, 1·2 kg (b) 4% **4.** (a) 1250 (b) 325, 185
5. (a) (i) 5·71 ohm (ii) 144 m (b) 0·06 **6.** (a) 9 × 10² (b) £17·61
(c) 0·0753 **7.** (a) 78 ohm (b) 1790 (c) 219 mm **8.** 5·01
9. (a) 0·7326 (b) (i) 24·0 (ii) 0·017 (iii) 0·04 **10.** (a) 6$\frac{3}{4}$ (b) (i) 69·53
(ii) 21·13 (c) 0·43 (d) $\frac{8}{25}$ **11.** (a) (i) 28·72 (ii) 0·0323 8 (b) 2·1 × 10²
12. 128 **13.** (a) (i) 46·4 (ii) 0·4488 (b) 7·2 **14.** (a) £28 (b) 3·2 × 10
15. (a) 1·951 (b) 23·2 **16.** (a) 0·23 (b) 25·54

Section B
17. (a) (i) $2a^{11}b^7$ (ii) $2ab^2$ (b) (i) $2a^2 - 3ab - 2b^2$ (ii) $a^2 - 2ab + b^2$
18. (a) $81x^4 y^{12} z^8$ (b) $x^2 + y^2 = 3xy$ (c) -5 **19.** (a) 1 (b) $\dfrac{2s - vt}{t}$ (c) $\frac{3}{2}$

20. (a) $\frac{1}{8}$ (b) $a = 0{\cdot}3, b = 0{\cdot}4$ **21.** (a) $y = -2$ (b) $36x - 84$
(c) (i) $2x(y + 2a - 3)$ (ii) $(x - 2)(3 - 4y)$ **22.** (a) $y = 3$ (b) $k = -14$

23. (a) $\dfrac{Pl}{R}$ (b) $\sqrt{\left(\dfrac{3V}{\pi h}\right)}$ (c) $\dfrac{V^2 - U^2}{2s}$ (d) $\dfrac{20}{20d - P}$ (e) $\dfrac{P - Wu}{W + P}$

24. (a) (i) $\frac{2}{3}b^2 c$ (ii) $9a^7 b^{11}$ (b) 4 **25.** (a) ay (b) $2a^2 + 5ab + 2b^2$ (c) (i) $a - b$
(ii) $2x + y$ (d) 3

26. (a) 25 (b) (i) $\dfrac{5\,ac}{2\,b^2}$ (ii) $\frac{1}{10}$ (c) (i) $\dfrac{y(2x - 1)}{2}$ (ii) $\dfrac{2pr}{4r - 3p}$

27. (a) $k = 4, t = -3$ (b) $P = 6, Q = 26$

Section C
28. $v = 11, t = 2{\cdot}35$ **29.** (i) 4 (ii) 2 **30.** 75·7 mm
31. (i) 95 mm (ii) 5·5 kg (iii) 64·5 mm **32.** 71 **33.** 8·3 A

Section D
34. 60 mm **35.** 75 mm **36.** (a) $45\frac{1}{2}$ cm (b) 6·7
37. YZ $= 12$, PY $= \frac{60}{13}$, PX $= \frac{25}{13}$, PZ $= \frac{144}{13}$ **38.** (a) 8·5 m (b) 8·7 m **39.** 2·8
40. 247 mm **41.** 45 mm

Section E
42. 12·0 m³ **43.** (a) 2290 mm² (b) 2·29 kg **44.** (a) 332 cm² (b) 321 cm³
45. (i) $1{\cdot}04 \times 10^4$ cm³ (ii) 4940 cm² **46.** (a) 1210 m³ (b) 113 kg
47. (a) Area $= \frac{1}{2}D^2$ (b) 56·0 cm² **48.** (a) 2·19 m (b) 790 m² **49.** (a) 2240
(b) 290 kg **50.** $5{\cdot}86 \times 10^{-2}$ m³ **51.** 26 700 **52.** (a) (i) 88·85 (ii) 18° 40′
(b) (i) 141 mm (ii) 20° 47′ **53.** 26·1 m **54.** (a) 351·5 mm (b) $\frac{3}{5}, \frac{4}{3}$
55. 90°, 7·0 m **56.** (i) 18° 24′ (ii) 55·0 mm **57.** (i) 35° (ii) 12·2 m **58.** 1·73 mm
59. $x = 50$ mm $y = 86{\cdot}6$ mm **60.** 25° 22′ **61.** 520 mm² **62.** 109·3 mm

63. 476 **64.** 27·3 kg **65.** (a) $\dfrac{5x^3}{y^2}, \dfrac{8}{5}$ (b) 11011 (c) 54

66. (a) 56°, 91°, 33° (b) (i) 840 cm³ (ii) 8·2 cm **67.** 25·1 cm

68. (a) 101001 (b) 100 (c) $\dfrac{a^3}{b^2 c}$ **69.** 120 m² **70.** 7·77 m²

71. (a) 37·5% (b) 27 100 mm² (c) 231 mm **72.** 141 mm **73.** 10 mm
74. 49·5 mm **75.** $x = 125$ mm, $y = 216{\cdot}5$ mm **76.** 221 mm **77.** 71·5 mm

Answers to assessment tests

Assessment test 1

1.

A	B	C	D
2	4	3	1

2. (a) y (b) x (c) z (d) z

3. 2, 5 **4.** 36, 3
5. (a) false (b) false (c) true (d) true (e) false (f) true

6. (a) **7.** (c) **8.**

A	B	C	D
4	2	1	3

9. 360·1, 7·6, 700, 900   **10.** 0·03684, 0·00964, 0·00841, 0·00803
11. $\frac{11}{12}$, $\frac{5}{6}$, $\frac{3}{4}$, $\frac{2}{3}$ **12.** (a) 2   (b) 15   (c) 7   (d) $\frac{10}{7}$, $\frac{4}{7}$ (e) 1

13.

A	B	C	D
2	3	4	1

14. 0·793, 0·821, 917, 136

15.

A	B	C	D
1	3	4	2

16.

A	B	C	D
4	1	2	3

17. 3, tens, 7, tenths, 5, thousandths   **18.** (b), (d)   **19.** No
20. (a) four   (b) five   (c) decimal places   (d) significant figures
21. (a) false   (b) true   (c) true   (d) false

Assessment test 2

1. 3 : 2   **2.** (d)   **3.**

A	B	C	D
3	1	4	2

4. 4 cm   **5.** 18 m²

6. 3 : 6 : 15   **7.** (a) 9^2 (b) 5^2 (c) $\frac{7}{10}$ **8.** 25 °C   **9.** 2

10.

A	B	C	D
3	1	4	2

11. (a) $\frac{1}{4}$ (b) $\frac{6}{25}$ **12.** (a) 0·14   (b) 0·67

13. (b)   **14.** 2%   **15.** 120   **16.** 7, 3, 3   **17.** 8^{-2}, 8

18. (a) 9, 5   (b) 7, 8   (c) 15, 4   (d) 0, 1   **19.**

A	B	C	D
2	4	1	3

20. (a) 7×10^3 (b) $7\cdot4004 \times 10^3$ **21.** (b)

Assessment test 3

1. (b)   **2.** (d)   **3.** (a)   **4.** (d)   **5.** (a)   **6.** (c), (d)   **7.** 1·7782
8. (b), (d)   **9.** 1·7   **10.** (a) $\frac{1}{300}$ (b) 13   (c) 60   (d) 1600
11. (a) added   (b) index   (c) one   (d) two   **12.** (c)
13. (a) 20   (b) 63·25   (c) 0·6325   (d) 0·2   **14.** (c)

15.

A	B	C	D
2	3	1	4

16. (b), (d)   **17.**

A	B	C	D
4	1	3	2

18. (a) 149·1   (b) 0·5099   (c) 663·1   (d) 0·05848   (e) 4·135
19. (a) 21·16   (b) 2·145   (c) 0·2174   **20.** (b), (d)

Assessment test 4

1. (a) 8   (b) t (c) 5   **2.** (b)   **3.** (a) +   (b) −   (c) +   (d) −

4.

A	B	C	D
2	4	1	3

5. (c) **6.** (a) + (b) − (c) − (d) +

7. $38a$ **8.** (a) 6, 5 (b) 2, 4 (c) 4, 8 (d) 1, 3 (e) 2, 3 (f) 1 (g) 2
9. (a), (c) **10.** (c), (d) **11.** (b), (c), (e)
12. (a) $4x^2$ (b) $8x^3$ (c) $\frac{3}{4}$ (d) $\frac{3}{5}$ **13.** (b) **14.**

A	B	C	D
2	3	1	4

15. (a) adding (b) divided (c) zero (d) $\frac{1}{2}$
16. (a) = (b) < (c) > (d) = **17.** (a) X (b) X (c) Y (d) Z
18. z **19.** (c)

Assessment test 5

1. (a) equation (b) expression (c) equation (d) expression
2. (a), (c) **3.** (a), (c) **4.** (b) **5.** (c) **6.** (a), (c), (d) **7.** (c)

8. $x = 6\frac{1}{2}$ $y = 3\frac{1}{2}$ **9.** (ii), (iii) **10.**

A	B	C	D
4	1	2	3

11. (a), (b)

12. (c) **13.** (a) 9 (b) 10 (c) 3 (d) 5 **14.**

A	B	C	D
2	4	1	3

15. (c)

Assessment test 6

1. 13, 22 **2.** (a) 12 (b) 4 (c) 16 (d) No **3.** −6, $2\frac{1}{2}$, 19, 15

5. (b) **6.**

A	B	C	D
2	4	1	3

7.

A	B	C	D
(d)	(a)	(b)	(c)

8. (b)

9. (c) **10.** (a) **11.** (c), (d) **12.** 25, 2

Assessment test 7

1.

A	B	C	D
3	4	1	2

2. (c)

3. classes, range, frequency, class: total. **4.** (d) **5.** (d)
6. 30%, 0·955–0·995, 0·995–1·035, 1·035–1·075 **7.** (c) **8.** (d) **9.** (b)
10. false, true, true, false **12.** (b) **13.** (b)

Assessment test 8

1. (a) rotation (b) 360° (c) 60° (d) 180° **2.**

A	B	C	D
2	4	1	3

3. (b)   **4.** (b), (c), (e)

5. (a) no, no   (b) two, two   (c) three, three   (d) hypotenuse

6.

A	B	C	D
4	1	2	3

7. (b)   **8.** (a) x (b) β (c) θ, y (d) α

9. (a) COD, AOF   (b) DOC, AOC   (c) AOF, FOE   **10.** (b), (c)

11. (b), (e)   **12.**

A	B	C	D
4	1	2	3

13. angles, proportional, equal

14. (d)   **16.** (d)   **17.** (b), (f)   **18.** (a) $\frac{20}{3}$ (b) 15   (c) 12   (d) $\sqrt{3}$
19. (a) 50°, 40°, 140°   (b) 120°, 60°, 60°, 60°   (c) 50°   (d) 60°   (e) 40°
20. (d)   **21.** (a) true   (b) false   (c) false   (d) true

22.

A	B	C	D
4	3	2	1

23. XY, BC, XZ, AC, YZ, AB

Assessment test 9

1. (a)   **2.** (c)   **3.** (f), (d)
4. 90°   (b) equal   (c) perpendicular   (d) half   **5.** (c)   **6.** (b)
7. (a) 90   (b) 28°   (c) 1·6   (d) 2·1678   **9.** (b)   **10.** (c)   **11.** (b)
12. (a) equal   (b) circle   (c) 120°   (d) equilateral
13. (a) 30°   (b) 5π (c) 45°   (d) $\frac{2}{3}\pi$

15.

A	B	C	D	E	F	G
6	4	3	7	5	1	2

16. (a) 54   (b) 62   (c) 70   (d) 30

17. (a) 4   (b) rhombus   (c) parallel and equal   (d) square
18. (a) quadrilateral   (b) trapezium   (c) parallelogram   (d) rhombus
   (e) rectangle   (f) rectangle   (g) square   (h) quadrilateral

Assessment test 10

1. Same area   **2.** (c)   **3.** (a)   **4.** (c)   **5.**

A	B	C	D
4	1	2	3

6. (c)   **7.** (b)   **8.**

A	B	C	D
2	4	1	3

9. (b)   **10.** (c)   **11.** (d)

12. (a) cylinder   (b) cone   (c) hollow cylinder   (d) sphere
13. (a) 1 cm³, π cm³   (b) 6 cm², 4π cm³   (c) $\frac{1}{6}$, $\frac{1}{4}$ **14.** (c)
15. (a) cone, cylinder, hemisphere   (b) cube, pyramid   (c) cone, cylinder
   (d) cylinder, hemisphere   **16.** $\frac{4}{3}\pi x^3$, sphere of radius x **17.** (b)
18. (b), (d)   **19.** false   **20.** (c)   **21.** (d)

1. (a) false (b) true (c) false (d) true (e) true

2. (a) $\dfrac{1}{\sqrt{2}}$ (b) 45° (c) 60° (d) 45°   **3.** (a) 3 (b) 1 (c) 5 (d) 2

5.

A	B	C	D
3	4	2	1

6.

A	B	C	D
4	3	1	2

7. (a) $\cos\theta$ (b) 1 (c) 360° (d) the curve repeats itself (e) -1

8. (a) 0·7819, 0·6234 (b) 49° 42′ (c) 0·8238   **9.** (a), (b)   **10.** (b), (d)

11. (a), (c), (d)   **12.** (c)

Index

Addition of fractions, 10
Aids to calculations, 55
Aids, comparison of, 73
Algebra:
  addition and subtraction, 79
  multiplication and division, 82
Angles:
  alternate, 159
  corresponding, 159
  properties, 157
  supplementary, 160
  types, 156
Angle at the circumference, 190
Angle of elevation and depression, 250
Angles of a triangle, 162
Annulus, area of, 218
Approximate values, 54
Arc, length of, 196
Area of:
  annulus, 218
  circle, 217
  parallelogram, 211
  rectangle, 211
  square, 210
  surface of cylinder, 226
  surface of prism, 224
  trapezium, 212
  triangle, 212
Area, units, 209
Associative law, 3, 18
Axes:
  Cartesian, 129
  parallel, 125
  scales, 130

Bar charts, 144, 146
Binary addition, 47
Binary numbers, 45
Binary to denary, 46
Binomials, 97
Brackets, 95

Calculators, 72
Cartesian axes, 129
Characteristic (logarithms), 60
Chords, 194
Circle:
  area of, 217
  parts of, 188
Circumference, 189
Commutative law, 3, 80
Cone, 232
Congruency, 170

Construction of:
  equations, 110
  right angles, 169
  triangles, 176
Co-ordinates, 133
Cube roots, 5
Cylinder, 226

Data, display of, 141
Decimals, 16
  addition and subtraction, 22
  multiplication and division, 22
  recurring, 21
  reduction to a number of places, 20
Decimals to fractions, 18
Denominator, 9
Directed numbers:
  addition and subtraction, 85
  multiplication and division, 89
Distributive law, 3, 81
Division of fractions, 14

Equations:
  construction of, 110
  simple, 106
  simple with brackets, 109
  simple with fractions, 109
  simultaneous, 112
  transposition, 116
Equations and expressions, 105
Errors, 53
Evaluation of formulae, 114

Factors:
  in algebra, 97
  in arithmetic, 4
Fractions, vulgar:
  addition and subtraction, 10
  improper, 9
  mixed, 9
  multiplication and division, 13
  proper, 8
  simplification, 9
Fractions to decimals, 19
Frequency, relative, 143
Frequency table, 141

Gradient, 134
Graphs:
  cosine, 256
  sine, 256
  straight line, 133
  values from, 134

Hexagon, 202
Highest Common Factor, 7
Histogram, 148

Indices:
    algebraic, 90
    arithmetic, 39
    fractional, 41, 94
    negative, 40, 93
    zero, 40, 92

Link lines, 125
Logarithms of:
    numbers greater than 1, 59
    numbers less than 1, 65
    powers and roots, 68, 69
    unity, 67
    use of tables, 59
Lowest Common Multiple, 8

Mantissa, 60
Mapping, 129
Multiplication of fractions, 13
Multiplying by tens, 17

Numerator, 9
Numbers directed:
    addition and subtraction, 85
    multiplication and division, 89

Origin, 129

Parallelogram, 201
Parallel axes, 125
Percentage, 35
Perimeter, 209
Pictogram, 146
Pictorial displays, 143
Pie chart, 144
Precedence, rules of, 1, 83
Prime factors, 4
Prism, 221
Proportion:
    direct, 32
    inverse, 33
Pyramid, 23
Pythagoras' theorem, 166

Quadrilateral, 200

Radians, 196
Radians to degrees, 198
Ratio, 29

Reciprocal tables, 57
Rectangle, 201
Rhombus, 201
Rules of precedence, 1, 83

Scales, 130
Significant figures, 21
Similar triangles, 173
Simple equations, 106
Simplification of fractions, 9
Simultaneous equations, 113
Slide rule, 71
Sphere, 229
Square, 203
Square root, 5
Square root tables, 55
Square tables, 57
Standard form, 43
Statistics, 140
Subtraction of fractions, 11
Substitution, 83

Tables:
    logarithm, 59
    reciprocal, 57
    square, 57
    square root, 55
    trigonometric, 243
Tangent, 193
Transposition, 117
Triangles:
    congruent, 170
    construction, 176
    similar, 173
    trigonometric solutions, 248
    types, 163
Trigonometric graphs:
    cosine, 255
    sine, 254
Trigonometric ratios, 242
Trigonometric ratios:
    30°, 60°, 45°, 246
Trigonometric waves, 256

Volume of:
    cone, 232
    cylinder, 226
    prism, 221
    pyramid, 231
    sphere, 229
    units of, 222
Vulgar fractions, 8